COAL-CAMP CHRONICLES

Coal-Camp Chronicles
Clyde Roy Pack

©2005

PUBLISHED BY WHERE? PRESS, INC.

PAINTSVILLE, KY

COVER ART BY THE AUTHOR

ISBN 0-9719144-5-1

No part of this publication may be reproduced in whole or in part, or stored in a retrieval system, or transmitted in any form or by any means, electronic, mechanical, photocopying, recording or otherwise, without written permission of the publisher. For information regarding permission, write to Where? Press, Inc. P.O. Box 154, Paintsville, KY 41240

DEDICATION

To the four most important people in my life: Wilma Jean, Todd, Marcy and Alison.

ACKNOWLEDGEMENTS

Special thanks to those who helped tell our story via photos: Dr. Joe Spears, Carolyn Lyons Daniels, Margaret Webb and Harkless O'Bryant.

Many thanks, also, to Bryan Auxier of Where? Press, for his interest in our little project. His patience and understanding, not to mention his design skills, are greatly appreciated.

And, of course, we must thank Dr. Linda Scott DeRosier for all her support throughout the years. It was she afterall, who convinced us that people still cared about the Muddy Branch we knew and loved as kids.

Thanks to my parents, Willie and Julia Pack, my six brothers and sisters (Ulysses, Hubert, Goldia, Ernest, Mary Jean and Joe) as well as all the other people who, over the years, inhabited the little community of which I've written. Collectively, they made Muddy Branch a special place for a kid to live in the 1940s and 1950s.

Thanks to Wilma Jean for all her support and advice. Without her, this book would never have happened.

And finally, a gazillion thanks to my daughter-in-law Marcy and her proofreading skills.

TABLE OF CONTENTS

Introduction 5

Chapter 1   Friends and Neighbors 13

Chapter 2   A Special Time 37

Chapter 3   At Work and Play 52

Chapter 4   Neighbors in the News 69

Chapter 5   Around the House 94

Chapter 6   Learnin' 111

Chapter 7   Rememberin' Stuff 140

Chapter 8   A Distinct Culture 163

Chapter 9   Church and Such 179

Chapter 10 Tragedy: at Home and at War 193

Chapter 11 Northeast, According to the *Herald* 202

About the Author 247

Index 249

INTRODUCTION

"Now keep your feet under them covers, Clyde Roy," my mother said as she tucked around my body the heavy quilt she'd made from denim squares of various shades that had once been overalls belonging to me, my little brother Joe, or maybe even one of our older brothers.

"You too, Joe," she said to my bed partner, who feigned sleep, throwing in a very realistic snore for effect. Nearly three years my junior, I knew he'd not even attempt sleep until at least the third or fourth, "You boys quieten down, now. Your daddy can't sleep," would come from Mom and Dad's bedroom, the doorway of which was not more than an arm's length from the head of our bed.

Minutes passed in total silence except for the strong November wind that seemed to be running laps around our tiny coal-company house, slapping each corner noisily as it passed.

"Is it gonna snow tonight, Clyde?" Joe suddenly asked in a half whisper.

"The radio said it might," I answered back as quietly as I could.

"Boys?" came Mom's voice from the other room.

"Is it going to snow in Jay-Pan tonight?" Joe asked, breaking the word into two heavier-than-need-be syllables.

"Lord, Son, I don't know. Why?"

"How about Germany?"

"How would I know that? And why do you wanna know about Jay-Pan and Germany, anyway?" I answered, mimicking his pronunciation.

"Cause that's where Ulysses and Hubert is. It snows there, don't it?"

"Boys!" Mom said, a little more forceful this time.

"I hope it snows all the way up to Toe-Joe's hind end," Joe said.

I got tickled at that, but muffled the laughter into my pillow hoping Mom couldn't hear it.

Things got quiet again and I thought of our two oldest brothers and doubted that either of them was even in Japan or Germany. But as far as Joe was concerned, those were the only two places where War World II was being fought.

Sometime during the night we were indeed treated to our first snowfall of the season. Depending upon which window we looked from, Joe and I figured the depth ranged anywhere from six to eight inches.

"Looks like we got about an inch," Mom said, much to our disappointment.

***

That winter night in the early 1940s was typical of the life I lived as the son of an Eastern Kentucky coal miner. During daylight hours I ran through the Muddy Branch coal camp, rode saplings to the ground in the nearby woods, and was completely oblivious of the community's history, or even that there was a world of grown-ups around me — other than my own immediate family, of course. My oldest brother, Ulysses, was 16 years my senior. Next in line was Hubert, who was 14 years older than me. Then, in two-year increments, came Goldia, Ernest and Mary Jean. I came along seven years after Mary Jean, with Joe nearly three years behind me.

But all I knew about the coal mining industry was that Daddy was a miner and went to work clean and came home dirty. Therefore, most of the things I remember and have written about coal-camp life as it was in the 1940s and 1950s have come — and will continue to come — from that perspective.

But with adulthood, and many hours of research, I've come to discover that this little coal-camp community (Thealka, a.k.a. Muddy Branch) also has a rather historic past. So while my childhood memories will continue to dominate my writing, I wish to use this book to look back at the history of the place and tell the grown-ups' side of the story, too. I'm convinced that their story parallels life in every rural community in the country at that time. Muddy Branch was indeed a microcosm of America herself.

Unfortunately, I now find that many of the adults who made the community their home at the time I was running loose — such as my own parents — are no longer with us. Many of the others who are still living are scattered throughout the country, many living with their children who moved to other states in search of gainful employment after the mines shut down.

Such is the fate of these one-time Eastern Kentucky miners who helped keep coal supplied to the steel mills in the north, and who later abandoned pick and shovel to join the millions of young men and women fighting for our freedom during two World Wars.

Unfortunately, few public records were kept regarding the contribution these everyday, ordinary folk made to society.

Clyde Roy Pack

However, the pages of *The Paintsville Herald*, founded in 1901, a short five years before the Northeast Coal Company began exploring for black gold in the hills of Muddy Branch, help tell the tale of the comings and goings of the community's citizens. Few of them ever gained the slightest degree of fame, but the mere fact of their being there has been hidden for nearly a century in the long-forgotten volumes of our local newspaper. It has been a joy to once again bring to life the achievements of the ordinary folk who were a part of this community.

Some of the news reported about the residents of Muddy Branch was good. Some was bad. But either way, as I continue to describe life during the 1940s and 1950s as seen through the eyes of a kid (as told by the man the kid grew up to be), perhaps entries from the yellowed, brittle pages of *The Herald* will add a new dimension to a people who, likely without their even knowing it, helped shape this community, state, and, yes, even the nation.

***

The Paintsville Herald, July 25, 1907 — *Ordinarily the big coal operators have but little interest in the mental and moral welfare of the people who work for them. The Northeast Coal Co., however, is an exception to the rule. This company prides* itself on the pleasant surroundings of those in its employ. Every house it has constructed for the miners is neat and comfortable. The company has just completed a school building on Muddy Branch second to none in Johnson County, outside of Paintsville. The company will employ teachers at its own expense and every employee sending children to this school will be urged to see that they are regular in attendance.

A fine church building has also been constructed and a Sunday school is being held every Sabbath day. Every child in the vicinity of Muddy Branch, and men and women, too, for that matter, are given every encouragement to be regular in attendance. In addition to the Sunday school, services are held by ministers of the different denominations. The Northeast Co. is also having a number of fresh water wells drilled in order to supply those living around its mines with good pure water. The sanitary condition is closely watched that contagious diseases may not get a start. Muddy Branch is not only a model town but it is a large town. Our people might be surprised by the statement, but it is a fact nevertheless that the population of Muddy Branch equals that of Paintsville. Keep your eye on Muddy Branch and the Northeast Coal Co., for business will be done there on a big scale, before long.

Clyde Roy Pack

***

The above article is one of many that appeared in *The Herald* before 1910. Either publisher Warren Meek was a true believer in the company or the company itself had hired a good PR person. Meek's articles, many of which I'll share in this book, were all extremely positive in nature. But even if they might be considered biased reports, at least they chronicle events pertaining to the history of the only place in the world where I was born and raised. Unfortunately, for various reasons, some copies of *The Herald* have disappeared from existence, perhaps causing gaps in a particular storyline.

I take no responsibility for misspelled names or errors in reporting of the events, some of them occurring nearly 100 years ago. What I have recorded is what was printed at the time.

References I make to any person, place, or thing from my past should be regarded as nothing more than memories. I will admit to changing some names here and there because the last thing I would want to do is cause embarrassment to certain individuals.

Hopefully, this book will tell the tale of a group of hardy, God-fearing people who occupied this community over the past century. It is my desire to convey their language, customs and strong faith so that others may know who they were and the contributions they made to our society.

Clyde Roy Pack

Chapter 1

Friends and Neighbors

No two days were alike for a kid growing up in an Eastern Kentucky coal camp during the 1940s. Families were generally large, which meant that there were always others my age with whom to explore the unknown.

Hopefully, the stories I've included in this book about the people, places and things that made daily life for a pre-teen an adventure will suffice to make one understand that times then were no more like times now than daylight is to dark.

Of course, that's how it should be. I mean, who would want to go back to paying 20 cents a gallon for gas or a dime to get into a movie? On the flip side of that coin, however, is the fact that a man working for as much as $10,000 a year was considered to be living high on the hog. Coal miners definitely were not.

Putting all that stuff aside, though, one of the things that really defined the times as much as anything else had nothing to do with one's financial status. The most memorable difference between then and now, at least as far as I'm concerned, was the closeness that existed among neighbors. That's not to say that every neighbor got along perfectly with every other neighbor. As a matter of fact, there were even a few good-natured feuds between families and at least one full-fledged one. But that's another story for another time. Feuding in general, though, was not the shooting kind like the Hatfields and the McCoys or anything like that; it was more like a one-upmanship thing, like whose crew would load the most coal or whose boy was the tallest for his age. Stuff like that.

But despite the fact that certain individuals were in constant competition, when the chips were down, everybody banded together. This was never more evident than in times of tragedy. Families that were said to never have "gotten along'" often stood side by side and supported one another 100 percent.

Furthermore, in those days one was your neighbor if he lived within a mile of you. Everybody knew everybody right down to the names of all their children, and in some cases, even their ages. And with every family having four or five kids, that took some doing.

Nowadays, you're lucky if you know the names of the folks next door, let alone take the time to stand a few minutes every once in a while and carry on a conversation.

In that respect, those of us who experienced life back then were extremely fortunate. We just didn't know it.

Clyde Roy Pack

As I mentioned in the introduction, as a coal-camp kid about the only thing I knew about the mining industry was that my daddy was a miner who went to work clean and came home dirty. What he did in the time he was away from home and what became of the coal he helped dig during that period was of absolutely no interest to me.

I'm not particularly proud of my ignorance about such matters, but neither am I about to beat myself up over it. After all, I was just a kid and such things as John L. Lewis and unions and tonnage and the price of coal just simply didn't interest kids.

Fast-forward half a century and after it's much too late to get any first-person reports on all that stuff, I find myself becoming curiouser and curiouser about the industry that put food on our table and eventually made my mother a widow. Fortunately, though, there is still available written documentation that can shed a great deal of light on the coal company that fed me, and how things really were back then.

The following is an example.

***

The Paintsville Herald, **May 12, 1927** — *A. Dw. Smith, president of the Northeast Coal Company, and an honorary member of the Paintsville-Van Lear Rotary Club, made a talk at* the club meeting Tuesday and gave the members some valuable information on the operations of his company in Johnson County.

He said that much interest was being shown over the country by a tunnel in Colorado that was nearing completion which was six miles long and handled both trains and automobiles, shortening the distance from coast to coast, but said right here in our own county we had a wonderful tunnel made by coal mining that was 820 miles long. This tunnel was made by the Northeast Coal Company alone, and does not include the other coal operations in the county. If this tunnel widened to that of a railroad tunnel, it would reach to Cincinnati. These were facts few of the members ever thought of. This shows, the speaker stated, what a little work at a time will accomplish.

***

In previous research, we've learned that Smith was not only the company's president, but was also one of its founders who was obviously proud of its achievements. And, if he was right about the number of tunnels in Johnson County — and we've no reason to doubt his word — when you consider all the other underground mining operations in Johnson County, the area just beneath our rugged landscape must surely look like a honeycomb.

But I suppose that truth be known, even if I had been aware of information such as this when I was a kid, I likely wouldn't have understood it anyway. Matter of fact, I'm not real sure I even grasp it now.

Perhaps it's an unintentional comparison of the way things were to the way things are, but it seems we're hearing an awful lot these days about the "greatest generation."

I don't know if Tom Brokaw actually coined the phrase by naming his book that, but the term sure seems to be used a lot, and from all indications, it more than adequately describes those who endured World War II, both as participants as well as those who stayed home. It appears to have been a time when the entire focus of the country was on the military, especially those men and women serving overseas. From what I've read, "Rosie the Riveter" was much more than just a cute little slogan.

So much for the "way things were." Unfortunately, we've come a long way from the days when I was just another of the thousands of coal-camp kids who helped populate Eastern Kentucky, and we now live in a time when selfishness has overtaken selflessness. Although there are still those among us who are as patriotic as anybody was during WWII, I'm afraid that far more Americans aren't, than are.

So what's happened? Why can't the government be trusted? Why is our average citizen seemingly filled with cynicism regarding anything and everything our leaders say and do? Talk to twenty people and you'll get twenty answers. Likely as not, they'll all be valid.

I was a bit too young to be considered a bonified member of the greatest generation, and about the only thing I can remember of the war was that Ulysses and Hubert were in the Navy and were in the Pacific Theater. I never heard a single complaint from either of them (or my parents, for that matter) about their having to be there.

As a matter of fact, I'd say that Muddy Branch as a whole more than met its quota when it came to providing soldiers. It'd be pretty safe to say that every set of parents (and parents in those days did indeed come in "sets") who had a son between 17 and 25 who wasn't physically disabled went though the same degree of prayerful worry as did my own.

The primary difference between the Muddy Branch I remember and now, however, is that back then everybody was on the same team. Now, it appears that about the only folks who support the military are those who have sons or daughters or grandchildren actually in the service. Everybody else does little more than gripe about our country's leaders.

Regardless of which side of the fence your opinions fall, guess we'll all have to agree that history will never, in a million years, be as kind to us as it was the generation that survived WWII ... the generation I write about in this book ... the generation under which I morphed from kid to adolescent to young adult.

***

It must have been about 1944 when we moved from the last house in Smoky Hollow to next to the last house in Silk Stocking Row, a distance of not more than half a mile. All I remember for sure about the change in our residence was that our large furniture — mostly beds, mattresses and the kitchen table — was moved in a big sled pulled by a mule driven by Earl VanHoose's boy Jargo. He made two or three trips and Dad paid him two dollars.

That's where we lived when I started school in early August 1945. We lived there until I was in the eighth grade. Then we moved again, right back up on Smoky, but not in the last house this time.

Actually, as far as I was concerned, Silk Stocking (a.k.a. Society Row) was the best place for a kid to live because there were so many other kids around to play with. Not that I didn't have a lot of friends when I lived up Smoky, too, like Keith Lyons and Jimmy Spencer.

As was typical of a coal camp, some families moved in and out of Silk Stocking rather quickly. Yet, names and faces of youngsters still linger in my memory. For example, John Burton's family lived in the very last house and he had two daughters (Katie and Donna) about the age of Joe and me. After they moved out, the Claude VanHoose family moved in. Their two sons, Paul and Wilbur, were about our age, too. The VanHooses had two older daughters, Claudine and Charlene (we called her Shod).

The house next to us on the lower side was occupied for a while by the George Reynolds family, who had an older daughter, Georgene (who married my oldest brother, Ulysses), and Patsy Grace, perhaps a year older than me. Theodore Miller and his wife Doris lived in the camp, too. Their son Mike was a bit younger than us. The Millers lived next door to Virgil Green's family. The Green kids our age were Jimmy, Paul and Libby Ann.

After the Greens moved away, John Daniels and his family moved in. They had P-Jack and Tiny. Milt Ratliff's family lived right below them and had Roger and Margaret Ann, about our age. They had two older sons, Paul and Fred. Jeff Sparks and Doll lived near them with their foster son Eck, and Lizzie Colvin lived a couple of houses away, with her grandson Tucker Daniel and daughter Lois Ann adding to the list of youngsters growing up in the community.

Of course, there were others who came and went at various times in the lower end of the camp. James O. and John Martin VanHoose, Billy Boy Bailey, Charles Lee Price, James Roger and Tom Castle, Norman and Larry Griffith, June Bug Nelson, and James Harvey Prince come to mind. Kids we played with from Greentown included Johnnie Trimble, Jack Cecil Lyons, Buck Malone and Nooner and Delmas Fitch.

I've no doubt omitted some, but suffice it to say, kids were plentiful in and around that particular coal camp and there seems to have developed a certain bond among those of us who proudly refer to ourselves as coal-camp kids. Apparently, our coal camp was just as dusty and exhibited the dull, monochromatic sameness as a coal camp in Pennsylvania, West

Virginia, or any place else where men mined coal in the 1940s and 1950s.

When we lived in those simple little box-like structures that we now nostalgically refer to as "company houses," we couldn't have imagined that we were living in what would become known as historic times; that the jobs our fathers did day in and day out were literally heating, lighting, and powering this entire country; that the steel used to build big-city skyscrapers (that they never saw) and span rivers (that they never crossed) existed because coal had been hauled to blast furnaces in the North from beneath the lush hills and out of the long, winding hollows of some of the most remote regions of the country.

That kind of thinking was far from the minds of the coal-camp kids I knew, however. It wasn't until years later that historians, while explaining much of the technology and machinery used in mining of this ancient fuel, also made us aware that there was a certain human element involved within the glorious, but dangerous history of coal production. Perhaps that's one of the things that has led to the feeling that we have indeed evolved from a special people.

Exactly how the Muddy Branch coal camp, located in rural Johnson County, Kentucky, evolved may never be known, but one can only imagine that it happened something like this:

Three wealthy entrepreneurs, the aforementioned A. Dw. Smith, Morris Williams and Carl Metzger, from Philadelphia, Pennsylvania, decided around the beginning of the last century that coal was in abundance in this part of the world. Holes of exploration (there's no record of exactly how many) began to dot the landscape. News went out via word-of-mouth and print ads to wherever men seeking work might be found. Consequently, soon-to-be miners and their families, from northern cities to southern farms to tiny towns in Europe, flocked to this area.

Needing to house the families, coal operators hastily began building the company houses. Pretty soon, railroad tracks snaked around hillsides and through the creek beds.

***

The Paintsville Herald, July 4, 1907 — *The grading will soon be completed to the head of Muddy Branch for the branch railroad. The Northeast Coal Co., putting in the road, is arranging to make a number of openings at the head of that branch.*

A tipple is to be constructed at the head of the branch, at the Jim Dills farm and openings made on the two hills there. Other openings will be made later.

The Northeast is still constructing houses on Muddy Branch. The big combination office and commissary building is well under way and a number of new residences for miners are being erected. The output at the mines is gradually on the increase. In fact, the company is experiencing trouble getting cars to load the amount of coal it is mining. The population on Muddy Branch is increasing, new miners moving in almost every day. Muddy Branch is the coming mining town of the Sandy Valley.

***

Schools were built and staffed and the company found a good man or two to preach the word, with the school building doing double duty as a meeting house. This of course, led to the construction of a church building, with the church's denomination likely decided by whatever religious beliefs were predominant. (In our case, make that dyed-in-the-wool Free Will Baptists).

Then, to keep coal-camp families from spreading their new-found wealth too far afield, before long the company store was built. Credit, initially in the form of scrip (coins stamped with the company's name and its value in currency) was issued to the miners and a percentage of what they purchased, along with the rent for the company house in which their family lived, was withheld from their pay at the end of every half. (Miners were paid every two weeks, or every "half" month.)

Thus the Muddy Branch coal camp was built and maintained by the Northeast Coal Company. But we coal-camp kids didn't know, and likely didn't care, about any of this. Yet, even though we weren't aware of it, we grew up with a feeling of camaraderie for others who grew up like us. And deep down, as time goes by, we become prouder and prouder of our heritage.

Anyway, when the company finally began burrowing beneath the hills of Muddy Branch, just as it was in coal camps throughout Appalachia, housing for families of the miners who would dig the coal was all part of the grand scheme of things. In our case, the fact that there wasn't enough level land for one house seat, let alone dozens, didn't deter the builders, however, and the dwellings were constructed — all the same size, all the same color — in crowded rows along the hillsides, creek beds and hollows surrounding the mining operations. Whenever a flat place could not be scraped out of the yellow dirt large enough to hold the tiny structure, many of them were built on stilts, usually facing the road, or what passed for one. Lots of times the back of the house would be on ground level and the front porch would be eight feet off the ground. The familiar Loretta Lynn homeplace in Van Lear's Butcher Holler — although built by another company — comes to mind.

However, stilts weren't necessary for the five-room, two-story houses in Silk Stocking Row. These houses were pretty much built on level ground.

(Having been arranged in such close proximity, not only was there a prevailing feeling of togetherness among the neighbors, but neither was it the least bit unusual for one next-door neighbor to be pretty much informed as to another next-door neighbor's business.)

And, with those thin walls, neither neighbor was necessarily required to be particularly nosey. All one had to do to get a front-row seat to a nocturnal yelling match between the couple next door (or a family fracas involving other members of the household) was to turn down the radio (remember, we didn't have TV then) and raise a window a little.

In the 1940s, I didn't know a single (here meaning one) woman who worked outside the home. So, the real beauty of coal-camp life as it regarded the houses being built so close together and all in a row, was that every back yard was fenced in (probably to keep livestock, which was plentiful, from wandering in) and each contained a well-worn bare spot right across the fence from another well-worn bare spot that came to be little points of rendezvous. If a wife had no particular business in her back yard and happened to look out the window and see a neighbor standing on her bare spot, it was a signal to her: "If you'll come out, I'll tell ya something."

Within an hour's time, an incident that had occurred in the last house in the hollow could be, and usually was, recounted in full detail, and usually with much embellishment, to the lady hanging out her wash fifteen houses away. It sort of reminds me of a little game we played in grade school where someone would whisper something to the student in the front seat of a row who in turn would whisper to the person behind him and so on until the person in the last seat in the row would get the message. Sometimes what filled the ears of the last listener varied greatly from the original.

Anyway, by the next day, even the men on their way to and from the mine were discussing the incident, or at least their wives' version of it.

"Say your old woman don't like your poker playin', Ed?"

"It ain't my poker, it's my drinkin.'"

But all-in-all, in those days when telephones were scarce, fence-line gossip was of no ill intent, and most of the time, harmless. Sometimes it even worked to the advantage of the infringed-upon parties. Anything out of the ordinary was usually reported, and it was not the least bit uncommon for a neighbor or two to show up at another neighbor's back door with a big cooker of soup beans or a chocolate pie if word had filtered down that the wife (or one of the children) was sick, or that the family was having a hard time because the husband had been laid off for a while. There was never a "How'd ya know I was feeling poorly?"

As strange as it might seem, more than sixty years after I knew the occupants of those houses, I can still picture many of them. And for some reason, it's the older folk I remember best. In some cases, they were indeed a colorful lot.

The famed Garrison Keillor (who is probably a bit younger than me) once wrote, "There's a survey out saying that people who take a positive view of aging actually live longer than those who grouse and grumble, which is hogwash and I am paying no attention to it."

Being familiar with Keillor via his books, CDs, tapes and occasionally seeing his "Prairie Home Companion" on PBS, I'm sure he intended his piece to evoke a smile or two. It did that, of course, but at the same time, based upon what I can remember about some of the older Muddy Branch citizenry I knew as a kid, his statement sure has a ring of truth to it.

It seems that many of the really old people I knew were as cranky as all get out, and the older they got, the crankier they became.

One person that comes to mind was Louella Scarberry (Aunt Lou to us kids), a pipe-smoking, hymn-humming, Bible-reading woman who actually seemed to get strength from her grousing and grumbling. She was really very kind to the kids who lived around her, but moped around like she was hardly able to put one foot in front of the other. That was until one of the neighborhood dogs would run through her begonias and she'd go on a tear of verbal abuse that would singe the hair off a terrapin's back. She'd then end up actually chasing the poor animal.

But she was, after all, a Christian lady, which is one of the things that now, more than half a century later, makes her so memorable, because the fun came from listening to her trying to cuss by substituting good, clean vocabulary for cuss words.

"You're as evil as old Hitler," she'd yell as she fanned the air with her cane. "You're worse than ... than tarnation, you sorry excuse for a ... for a ... thang, you!"

She left no doubt in the mind of any observer who witnessed her apron-flying, cane-waving tirade that had she been able to strike the animal, it would have been a goner. Everybody also understood that in her estimation, being a "thang" was a terrible thing to be.

But the point here is, maybe Keillor was right; maybe older folks even today need to be grouchy. After this dog-chasing lady would have such an encounter, for hours she'd be as perky and spry as someone half her age. It was almost as if the fit she had taken on the dog had pumped her full of adrenaline.

She lived well into her 90s and I attended her funeral at the Thealka Free Will Baptist Church. Ironically, her grousing and grumbling were not mentioned. According to the preacher, Aunt Lou had never uttered an ill syllable to anything or anybody. But I knew better.

Geographically speaking, Muddy Branch (or Thealka) is less than two miles northeast of Paintsville, the Johnson County seat.

The closest big towns are Huntington, West Virginia, about 60 miles to the northeast, and Lexington, Kentucky, which is about twice that far to the northwest. Thealka is still on the map but the place itself, at least as I knew it, has been reduced to memory. Of course the Thealka I remember existed near the end of the coal boom and was a far cry from what it had been 20 or 30 years before I was born. As I've stated, I knew little about the labor problems that accompanied the founding of the community and it wasn't until I reached adulthood that I was aware of a spectacular murder trial (documented later in this book by news accounts from *The Paintsville Herald*) that occurred in 1923 where several men were charged with causing an explosion at the Northeast mine that killed two individuals.

Anyway, about the only thing left of this once prosperous community are a few remodeled company houses ... and memories stored inside the heads of people like me. The word *prosperous* used in the previous sentence, although selected carefully, may raise a few eyebrows of those who were never fortunate enough to enjoy coal-camp life. Likely as not, to those folks the word conjures up thoughts of money. On the count of three, the rest of us can have a hearty laugh at that one. About the only money a coal miner ever saw was only for an hour or so at the end of the half when he'd open that tiny brown envelope and shake out the change left over after the company had held out a large percentage of already meager wages for rent of the company house and the groceries purchased at the company store.

The word "fortunate" (as used in the previous paragraph) was also used on purpose because the advantages gained from coal-camp living over a lifetime have more than served us well. Like today's pre-packaged hot oatmeal, prosperity comes in a variety of flavors and those who rubbed elbows with all the assorted personalities that made up the community have developed into a people who are tactful, tolerant and tough.

With miners coming to Appalachia from all over the world, Thealka residents, just as in other early coal camps, also represented a diverse ethnicity. But once on site, a neighbor was a neighbor and it mattered little to anybody from whence they came. About the only difference anybody really noticed, anyway, was whether or not the head of the family worked the day shift, the night shift or the hoot owl. Dad worked all three at one time or another, and while he never complained or gave any indication as to which shift he preferred, I liked it best when he worked the night shift. The reason, of course, was that I didn't have to rush off to bed with the chickens or sit and read quietly because he had to go to bed so early when he was on the day shift. When he worked on the night shift, however, it was like three in the afternoon until eleven o'clock at night and we could go to bed any time we wanted. Then Dad could sleep until way up in the day while we were at school.

Another thing that made a coal camp special was that there were no class distinctions. Everybody lived in a company house and it didn't matter whether you were a "boss" or had the dirtiest job in the mine, you were of equal rank inside the community. Everybody who could showed up at the Friday night pie supper being held at the grade school to raise money for a new lunch room. Even those on the night shift were in attendance for the Labor Day picnic, as cheers went up sporadically when Pick Colvin, or some other local Joe Dimaggio, connected for a homer at the baseball field, or new horseshoe champions were crowned amid the clang of metal on metal and the laughter of happy children filtering from the only area of abundant shade as provided by a huge sycamore.

The Northeast Coal Company store is gone now, having burned to the ground on February 5, 1957.

The old H. S. Howes Community School is gone, too. The company houses still there are no longer painted the once-familiar monochromatic sameness and are now equipped with cable TV (or pizza-pan-size satellite dishes), indoor plumbing, and tons of memories of a time when a sense of community was predominant and neighbors got along.

You didn't have to be an archeologist, however, to have figured out that the community did indeed have a past; that there had been a giant structure sitting on those large, box-like concrete shapes — many resembling some sort of royal throne like I'd seen in the movies — now protruding silently from the weed-covered hillside on both sides of the little gravel path. But back in 1949, as a 10-year-old on my way to the swimming hole on a sweltering July afternoon, I had no way of knowing just what that structure might have been, or how long ago it had been there.

It was many years later before I learned that the land the rag-tag group of coal-camp kids crossed on our way to another round of underwater tag had been the site of the Northeast Coal Company's Number One Mine and tipple.

Exactly when the million-dollar operation had been dismantled is as much a mystery now as it was then, except now I know that what I'd perceived to have been ancient ruins weren't really ancient at all. Since the company had not been founded until the early 1900s and the century then was yet to have reached the midway point, the demise of what surely must have been a noisy center of activity had not occurred simultaneously with the construction of the pyramids in Egypt. But, to a 10-year-old, whether it's 30 or 3,000, when you're speaking of years, they're both a long, long time ago.

We called the barrier that held the water in which we swam a slate dump, and if we had given any thought to it at all, it would likely have been that it was a natural formation instead of the result of years and years of many faceless, nameless men — long before our own daddies had been old enough to work — separating the waste from the coal and dumping it there, one tiny hand-pushed cart at a time.

Northeast's Number Three Mine still had half a dozen more years of life left in it and my dad and older brother Ulysses worked there. But, at the time, I had no idea there had ever been a Number Two Mine in Muddy Branch. The only thing I know for sure now is that it was located at Concord. I've been told that the tipple ran a conveyer belt over the C & O Railroad tracks and dumped the coal into barges that navigated the Levisa Fork of the Big Sandy. Of course, while I was discovering what was left of the Number One Mine, Number Two was also many years gone.

I was told recently that Northeast also operated a Number Four Mine on the Tutor Key side of the hill, with an entrance near where Jerry Daniels' store is now.

But, as would be imagined, back in 1949 as we moved single file toward the largest body of water any of us had ever seen, walking on tufts of grass whenever available and dodging sandbriars in an effort to protect our shoeless feet, we gave no thought to our history.

Strange, isn't it, how before we give thought to any sort of ancient ruin, we must first become one.

Chapter 2
A Special Time

Although I didn't enter the picture until 1939 and my memories didn't kick in until several years after that, a bit of research after I reached adulthood indicates that, in truth, I was born about ten years too late to reap the full benefits of coal-camp living. One-time residents (all a bit older than me) of the tiny dot-on-the-map community to which I lay claim as being my "hometown," have filled me in on some stuff of which I was just simply not aware.

For instance, I learned the company store once served more than its expected purpose. More particularly, its front porch had a multiple function. While that porch was merely a haven of rest for me on long, hot summer afternoons; a place to sit and eat chilly imps; to listen to older men brag of their coal-loading accomplishments in the Number Three Mine, I now learn it was also, just a few years before my arrival, a public arena; our version of Madison Square Garden upon which grown men, and sometimes grown women, settled their differences.

Perhaps it was because the combatants desired an audience (and that porch was plenty big enough to accommodate a sizeable crowd) or maybe it was simply the place where the disgruntled pair happened to meet — which was more likely, since the store provided every staple a family needed and was the only place this side of town that did — but for whatever reason, serious fights apparently occurred there on a regular basis.

Anyway, the reporter reporting to me said that on one particular January morning that was "colder than a well-digger's butt in Alaska," (his words, not mine) conflict between two grown women (one a miner's wife, the other the same miner's alleged girlfriend) was one of the wildest knock-down, drag-out events to ever be witnessed in Muddy Branch. He emphasized that since few women wore slacks in those days, everybody "got their picture taken." (Again his words, not mine).

He said there was no report of injuries beyond a few bruises and the loss of a couple of handfuls of hair to the two duking it out, but several innocent observers sustained minor cuts and bruises as the crowd was jostled back and forth as they made room for the two women who clawed and scratched and rolled from one end of the porch to the other. One woman (a non-combatant) who later claimed to have only come to the store for a box of kitchen matches and a twist of Brown Mule for her mother-in-law, found herself trying to walk on air when a sudden surge of human flesh pushed her from the high end of the porch (a good 10-foot drop) onto the railroad tracks. She escaped serious injury only because a tall snow bank had formed against some loaded coal gons that had been there but had been hauled away the previous afternoon to the Paintsville train yard by a C & O shifter.

So, it seems that as exciting as it was for my generation to have enjoyed coal-camp living, it was just plain tame compared to what it was like for the group of kids who preceded me by a few years.

I mean, of all the fights I witnessed as a kid, I never saw one where two adults were doing the fighting, and never witnessed anything on the company store porch other than a few miners chewing, whittling and spitting.

I'll admit, though, it was a good place to sit, even if you didn't chew or whittle and just wanted to count coal trucks.

"That's twelve," I'd say as a loaded truck lumbered by, followed by a cloud of thick, black dust.

The driver would get in line behind the three or four trucks waiting at the bottom of the little slope where a narrow road had been scraped from the yellow hillside that led to the tipple's hopper. He'd dump his load so it could be processed and loaded into the C & O coal gons that had been parked beneath the massive tin structure.

Within a minute or two, an empty truck would whiz by in the opposite direction, again stirring the black dust into a frenzy as it sped to wherever it was that it picked up its load.

On those long summer afternoons as I'd sit on the company store steps and eat a chilly imp and count trucks, I had no interest in where the coal came from. All I knew was that some of the empty trucks would turn left onto Route 581 and head in the direction of Tutor Key, while others would turn right and head toward town.

I'm not sure that at the time I was even aware that there were three sets of railroad tracks that ran beneath the tipple. One was for gons to be loaded with large lumps of coal, another was for medium lumps and the third was for fine coal. After it was dumped, the coal was carried by a belt to noisy screens that sifted the various sizes and when one gon was full, they'd push it out of the way and drop another one in its place. Every afternoon the C & O shifter would come and collect the loaded gons and pull them off to a train yard in town. I never gave a thought to where they went from there.

Sometimes I'd stop what I was doing in the late afternoon and watch the empty gons being pushed up a fourth set of tracks where 15 or 20 would sit waiting for the time they'd be needed. The trackful of empties sat on a pretty good grade, with gravity being the primary power that moved them to beneath the tipple.

Anyway, back to the truck counting, on days I had nothing else to do — that is to say if no one was particularly interested in going swimming or something — and depending upon who joined me in the cool shade of the store's front porch, sometimes I'd sit long enough to count as many as 75 or 80 loaded trucks.

But, I guess it's pretty obvious that we're a lifetime away from the days when the primary purpose of coal trucks was to give a kid something to count other that the number of days remaining before school started again.

***

Although it was no more than a half mile from our front door to the Thealka Free Will Baptist Church, in the summer time if we walked the main road when we went to Sunday School (and we always walked the main road because if we took the shortcut behind Foster Burton's house the briars would have snagged the legs of our good pants) by the time we got there, those shoes that Mom had insisted we polish — or at least wipe clean with

Dad's shine rag — would be covered with the thick black coal dust that covered everything else in the camp. It was the same dust that turned our bed linens to a pale shade of gray when Mom would dry them on the clothesline that Dad had strung between two poles in our back yard. It was the same dust that formed a thick film over every window pane in every house that always left ugly streaks whenever it rained.

Passing cars, and, of course, the coal trucks, even those driven at moderate speeds, would raise so much dust that we'd have to hold our breath until it settled. If we happened to play hard enough to raise a sweat, the dust would form little black rings around our necks like tiny, glistening pieces of black thread.

But when the autumn rains came, followed by those never-ending winter months — months that even Christmas couldn't brighten for very long — that thick, black dust turned into thick, black paste that clung to everything it touched, and it seemed to touch everything.

No matter how long you stomped and scraped your feet before you entered the house, you carried it inside. I've seen Mom sweep a handful of chalky, gray substance from her linoleum-covered kitchen floor after we'd walk through it just one time. She'd half-heartedly fuss at us for not wiping our feet, but I think she knew we could have stood out there wiping until the cows came home and still would have tracked it into the house. Even after it had dried the next morning, we'd take the shoes out on the back porch and slap them together and still not get all the caked mud off.

Coal-camp dust, which with the change of seasons would turn into coal-camp mud, was simply a way of life; something that was as much a part of the camp as was the constant roar of the nearby tipple or the abundant supply of neighborhood dogs that yelped incessantly as they chased every passing vehicle.

Imagine our delight when one summer day men from the county came with big trucks filled with hot asphalt and covered the main road with a thick layer of blacktop. Potholes once filled with red dog vanished and the dust that once chased the passing cars simply went away. Oh, there was still a dark cloud that hung heavily in the vicinity of the tipple, and everything white that Mom hung out on the clothesline still turned gray, but when the first blacktop road snaked its way from Route 581 all the way to the church house, it was a pretty big deal.

Of course, while you're experiencing it, you don't realize how very special the times are. I guess it just takes a heap of living before that happens. For example, it was many, many years before I came to understand that one of the many advantages of living in a coal camp was that we coal-camp kids had our very own railroad. Quite naturally, we also had an unwritten agreement with C & O and Northeast that allowed them to use it whenever they thought it was necessary.

No steam engines were ever on the tracks that led to the tipple located between the company store and the mouth of Silk Stocking Row, except when the small shifter belching thick, black smoke would push 12 or 15 empty coal gons onto the little siding so they could be "dropped down" and loaded at the tipple the next work day.

The string of empties ran well beyond the Free Will Baptist Church and since the railroad track ran between the road and the church, men from Northeast would break the chain of cars so the worshippers could have access to the building without having to climb through or crawl under the gons.

Anyway, despite warnings from our parents, we'd climb all over those empty gons, mainly, I suppose, because they were there and had a ladder going up the side of each one; and, we'd use their blackness to write messages on the sides of the gons with chalk we'd pilfered, one or two sticks at a time, from the chalk trays at school.

But one incident that occurred regarding those empty coal gons had nothing to do with climbing or writing. Instead, it dealt with the practice of placing pennies on the T-rails to be flattened. Or course, flattened, misshapened pennies were absolutely useless and the following story is the only thing of value that I've ever known to come from the practice.

In case you're not familiar with the art of penny flattening, what we'd do is take a penny and place it in the middle of the T-rail, right in front of the first wheel on the coal car. Then, when someone who worked at the tipple would walk up and drop the empty down, by the time several tons had passed over the penny, it'd be flat as a flitter.

One day as we stood by watching the process, as soon as the empty gon had passed, one of the smaller boys (we'll call him Frankie; you know, the change-the-name-to-protect-the-guilty thing) ran over and picked up the penny. It was a typically, hot summer day, and the penny was so hot when Frankie picked it up, it burned his hand.

Being a rather smart boy, he quickly threw it down. It was no big deal, really ... that is until Cecil (not his real name, either), one of the older boys, said, "Let's see your finger where it burned."

Frankie turned over his hand so everybody could see, and Cecil said, "Oh no! You picked it up heads."

The rest of us, not really knowing where he was going with this, but at the same time knowing Cecil, sensed it was going somewhere and wanted to be part of the ride, joined in with various degrees of our own sympathetic, "Oh, no!"

"What? What?" Frankie asked, with more than a little concern in his voice.

"Oh, nothing," Cece said. "You'll be all right as long as it don't blister Abe Lincoln's face on your hand."

"What if it does?" three or four of us asked in unison, hoping for at least a clue as to where all this was heading.

"You don't want to know," Cece answered.

By that time, Frankie was blowing frantically on the palm of his hand, which by now was hardly red enough to discern where the penny had burned him.

"I think it's gonna blister," he said. "I better go home."

He did, but in an hour or so, he rejoined the group who had now moved to the front steps of the company store. His injured hand — which had evidently worsened dramatically between the time he left us and the time he'd reached home — was heavily bandaged and reeking of the familiar smell of White Cloverine Brand Salve.

"You all right, Frankie?" one of us asked.

"Yeah, Mommy said she couldn't see Abe Lincoln."

"Good," said Cecil. "Maybe your fingers won't rot off then."

Frankie went home again.

Of course, Cecil and Frankie were not their real names, but that's not important because had there been a *Guiness Book of Records* and an entry for communities that bestowed nicknames upon its residents, the Muddy Branch in which I grew up would no doubt have been at the top of the list.

I could never begin to imagine where all those names came from. I mean, why would you call someone Fer Pole? Or Jargo? Or Three Prong?

Now, you can almost figure that if someone's head looked like Dumbo the Elephant's, he'd be called Flitter Ears. Or, if he were extremely short, he'd be called Mutt. A real skinny, tall boy might even logically be called Bird Legs. But, No-Not? And Pea Rat? How'd they ever come up with names like that? Apparently, some of those names were put on with Super Glue, for I've known some of these people for as long as I've known anything, and had it not been for a phone call from Jim Daniels after I had written about nicknames for my newspaper column ("Poison Oak") I would not have known Bud Meade's real name. Nor, Tiny Daniels,' and I just recently learned what Flop Smith's real name is.

I once heard someone say that only folks who are well liked are given nicknames. (Perhaps that's why I never had one.) Anyway, if there's any truth to that, most everyone else in the neighborhood must have really gotten along back in those days.

Anyway, about ten years ago, a bunch of us got together and tried to think of as many nicknames as we could from our younger days, and we gave a prize at the Muddy Branch Reunion that year for the person who could match the most given names with the made-up ones. I don't recall who won, but do remember the winner only got about half of them right.

We came up with quite a few nicknames, both male and female. Many of them are no longer with us, but listed alphabetically, here's who we remembered. We had an Auger, a Babe, a Babs and a Bake. Then, there was Beaver and Biddie and Bill Horse and Blackberry and Blue Head and Boe (could have been spelled Beau).

Who could ever forget Booten or Bott or Bozo or Brother? Although it's doubtful that many of us ever carried that much money on a regular basis, we also had at least four Bucks. We also thought of Buckwheat, Buddy, Bunk, Burr, Buss, Cat and Chig.

Then there were Coots, Cracker Bill, Crip, Crit, Crotchie, Dago, Darb, Dave-O and Dead Man.

We came up with names like Dickie, Dock, Dobbin, Doll, Dusty and Eck. Who could ever forget Fall Beans, Fatso, Fu-Nell, Good Nature, Grump, Hank or Hen-Sock?

Remember Hoodley, Hoot, Hoover, Hut, Jammer, Jim Paw, Jolly John, Lard and Leatherwood? We had us a Lang, a Long Jaw, a Moose, a Nooner, a P-Jack and a Paddle.

Wonder why they were called Peck or Perk or Polkberry or Puss? Rat was a preacher, Road wasn't. Moving right along, we had Rook Bill, Rooster, Scorpion, Shade, Shakey, Shod, Shorty, Slats, Slickum, Slue Foot, Smookem, and Squirrel. Tater Bug, Thacker, Thaw, Theadie, Tip, Toad, Toddy, Tooter, Trapper, and of course, Tucker.

There's no doubt that many an interesting tale could be told about the naming of Tug, Valentine, Wassie and White Eye.

For every nickname we remembered, there are probably three or four we didn't. Anyway, folks like Wib, Widdie, Wiggle, Windy and Zeal are still remembered with fondness and will forever be a part of the fading memories of my growing up in Muddy Branch.

***

We'd no doubt have called them Tubby or Pudge, but I can't think of a single kid in our neighborhood who was obese. Unfortunately, I can't say that about those same individuals some 50 years later because a dozen or so of us are currently carrying an extra good-size person around with us. But when we were kids? No way!

I guess it's pretty easy to figure out why that was the case among our young people, if you compare the way we coal-camp kids lived then with the way folks live now.

I mean, take sports, for example. When we played baseball, we really did play baseball. Like, we really did swing a bat, we'd really throw a ball (often water soaked and covered with black miner's tape), and we'd really run. Today, kids play sports with their thumbs as they sit in front of a TV screen with controls of some expensive video game.

If we wanted to go to the show on Saturday morning — only a couple of miles as the crow flies, but on those hot, muggy mornings seeming like twice that far — we walked. When nothing exciting was happening in the neighborhood, those of us who had one, would ride our bicycle with no particular place to go. At other times, we'd spend entire afternoons swimming in the Number One Pond in the head of Pond Hollow, or we'd hit the hills, fighting the briars and brambles, often ending up on the highest knob overlooking the camp.

In short, we got plenty of exercise. Add to that the fact that after school was out for the summer, when we'd get up in the mornings, our moms would shoo us out the front door, and even though they'd never say it, didn't particularly want to see us again until we came home about noon for a baloney sandwich and a big glass of Kool-Aid. Then, off we'd go again until suppertime. After supper, out we'd go again until darkness drove us indoors. Sometimes, even after it got dark we'd play a rousing game of kick-the-can or fox and hound.

We couldn't have gained weight even if we'd wanted to.

Chapter 3
At Work and Play

It's not likely that Dad, as proud as he was of the backbreaking work he was doing, or any of the other Muddy Branch miners, for that matter, had any idea of their contribution to society during World War II. It's too bad they're not still around to hear that historians have praised them highly and long since concluded that without them, and the thousands of others like them in other parts of the country, the outcome of the war may have been different.

But as the following story, gleaned from the pages of *The Paintsville Herald* indicates, even though the miners themselves may have taken their contribution to the war effort as trivial, at least others did not.

The Paintsville Herald, September 30, 1943 — *Miners in Eastern Kentucky will be urged to work a full six-day week, curtail absenteeism and step up coal production for the war effort, it was agreed at a joint meeting of United Mine Workers and operators' representatives in Lexington last week.*

Men who have been trained as miners and who have gone to other industry will be urged to return to the mines, and new men will be urged to take training for mine work, it was resolved in the meeting.

The miners' representatives were Sam Cady, Tom Raney, Edgar Reynolds and Ed Beane. Operator representatives were H. S. Horman, George S. Ward, A. E. Silcott and J. J. Ardigo.

The committee stressed the need for sufficient production of coal for maintaining the present high level of war production, and further for keeping the American people warm this winter.

We recognize, the committee announced, that there is a shortage of coal miners; that absenteeism for different reasons is more serious in the mines than in any other American industry. We recognize the fact that many men have left the mines for other industry and that very few people are being trained in the knowledge of mining.

The representatives resolved: That we will use all of our energy and influence, and that we will call upon all those whom we represent to use their energy and influence to correct this situation by undertaking to have all men who are working in the mines to work a full six days.

***

The old carpenter shop owned and operated by the Northeast Coal Company sat next to the tipple, on the right side of the railroad tracks as we walked toward town. Located about halfway between the company store and the mouth of Silk Stocking Row, the long, corrugated tin building, at least on work days, was always a hive of activity. The clanging of hammers and saws, metal striking metal, and the humming of strange-sounding machinery sang an unfamiliar refrain to young passersby on the way to and from school.

Kids were absolutely forbidden to even approach the broad front doors of the building when the men were working. Either Lonnie Castle or Garfield Stambaugh — two of the dozens of men who worked there, but the only two who still reside in the far recesses of my memory — would shoo us away if we strayed too close. Consequently, I never once saw the inside of the carpenter shop, except through tiny, coal dust-covered windows, perhaps on a Sunday afternoon when curiosity would get the best of me as I walked home from the show. Of course, on those occasions I saw very little except large ominous shapes silhouetted in semi-darkness against the windows on the far end of the building.

But the good thing about the old Northeast carpenter shop was its location. The back side of the building was completely out of sight from anywhere else in the camp. Therefore, the area in back of the shop became a gathering place on weekends for those old mean boys who played poker. Especially in warm weather, three or four of the older teens would spread large bandanas on the ground and hunker down to an afternoon of five-card stud or draw. When one would run out of pennies, nickels and dimes, he'd pour out his cigarettes and ante up with Luckies, Camels or Chesterfields. I don't remember — if I ever knew in the first place — how the worth of a single cigarette was determined. A whole pack cost about 20 cents so they couldn't have been worth more than a penny apiece.

Anyway, being too young to participate in such an illicit affair, once in a while I'd mosey back there just to watch. However, my little brother Joe put a stop to that one day when Mom heard him repeat a colorful phrase or two that he had picked up when he accompanied me to one of the poker games. He didn't get into trouble, but I did. I don't think I ever went back.

So, the old Northeast carpenter shop not only served its useful and expected purpose, it also provided a place of rendezvous for those seeking their fortunes — or at least, show fare — from the hands of Lady Luck.

Best of all, though, nearly 60 years after the lights in the old tin building were turned off for the last time, just its having been there has provided still another memory of a time when I'd never been so far from home that I couldn't get back before dark.

***

For some reason, it seems as if all my boyhood memories tend to be centered around the time I was about ten years old. It's almost as if I were ten years old until I woke up one morning and had turned eighteen. But I suppose age ten is as good an age as any since I don't claim that my accounts of how things used to be are historically correct anyway. However, those items lifted from old issues of *The Paintsville Herald* are likely correct, except for instances when long-ago reporters might have misspelled a name or overlooked an important detail or two.

Anyway, in the summer of 1949 many of the coal-camp kids who called Silk Stocking Row home would spend our mornings on or around our old familiar gathering place, the pump rock in front of Virgil Green's house. We didn't assemble to play games, though; that would come later in the day. What we did

Clyde Roy Pack

This happy little tow-headed feller is me, about seven years old.

was sort of mill around making small talk — somebody's cat had seven kittens; is Little Beaver a real Indian? — always speaking in soft tones as we kept our eyes and ears peeled.

That was the summer that Virgil's boy Ernie, who had been in the service and had learned to fly, would, on most mornings somtime between nine and eleven, come swooping over the ridge near Red Jacket Rocks and buzz the row of two-story company houses a couple of times, coming so close to the chimney tops that you could actually see his face when he waved. Once he even carried a piece of tree top on one of his wheels.

It was general knowledge (does that make it true?) that the red-nosed, yellow-winged craft he flew was a Piper Cub that belonged to Doc Turner. (John W. Turner, a Paintsville physician, reportedly kept the plane inside an old barn that he used as a hangar down on Bob's Branch in Thelma.)

But that was all beside the point because our focus was on the fearless pilot, braver, at least as far as were concerned, than Steve Canyon in the funny papers or Terry and the Pirates in our big little books. Besides, the man flying the plane was a real flesh-and-blood neighbor who lived only two doors away and whose brothers Jimmy and Paul, along with little sister Libby

Ann, were always among those in our imagined grandstand watching the best flying circus in town -- a show judged better than a Saturday double feature any old day. However, since I don't remember his ever flying on Saturday, we never really had a chance to put that last statement to the test.

Once, Ernie threw his cap out the window as he zoomed by, which, quite naturally, caused a mad scramble to retrieve it. Paul VanHoose claimed the prize and proudly handed it over to Ernie's mom, Bernice, also an interested observer of these aerial antics.

Although I didn't witness it first hand, local legend had it that Ernie Green — on more than one occasion — flew that plane beneath the swinging bridge that stretched across the Levisa Fork of the Big Sandy River and separated the front and back nines of the Paintsville golf course. The incident was much discussed that summer, especially among the boys who, to a man, decided that someday we too would learn to fly, borrow Doc Turner's plane, and entertain other boys and girls on some future muggy summer morning.

***

Another thing we often discussed was our someday learning to smoke. As far as I was concerned, that had to be the absolute coolest thing ever. And along with everything else the small community could ever need, the Northeast Coal Company's store sold Camel, Lucky Strike, Chesterfield and Old Gold cigarettes. The packs were stacked neatly behind a big wooden counter and two boys I knew, one about eight, the other about ten — and both to remain nameless — caught the clerk looking the other way and pilfered a pack of Luckies. They knew they needed some privacy if they were going to smoke in peace, so they got a couple of those big kitchen matches and headed for the old dairy barn in the head of Society Row. They each lit up one and had themselves a nice, leisurely smoke ... leisurely, that is, if you don't count the hacking and coughing.

But unbeknownst to them, they weren't as sneaky as they thought they were and the store clerk told their daddy what they'd done.

The next day on the way home from work, their daddy stopped by the company store and bought two packs of Luckies. He never said a word about it, but after supper that night, he called the boys out onto the back porch, sat them down and tossed them each a big brand-new pack of Lucky Strike cigarettes.

"Light up, boys," he said.

"But ... but you know we don't smoke, Daddy," one of them stammered.

"Oh, really? I heard you did," the old miner answered.

With a goodly amount of coughing and hacking, the boys finished a cigarette. As they were ready to toss it away, the father said, "Wait, light up again, right off that one."

That went on through about half a dozen cigarettes and both boys were turning green at the gills.

"Don't quit now," the father said. "You've not even smoked half a pack yet. If you're old enough to steal 'em, you're old enough to smoke 'em."

I don't know how many they smoked, but I do know for sure that both boys grew into manhood and to this day, more than 50 years later, neither of them has ever smoked another cigarette.

***

I doubt I so much as ever heard the word *linguist* until I found myself majoring in English at Eastern Kentucky State College in the late 1950s. Therefore, when I was growing up, I wouldn't have known a linguist from a load of coal. But for some reason, I've always wondered about some of the expressions I heard when I was a kid. But surely I'm not the only person who ever really wondered about "playing hob" and this Pat character that went to the army.

Over the years, I bet I asked every grown person I ever knew about these two things, and if I ever got an answer that made a lick of sense from a single one of them, I can't remember it. But sure as anything, when I'd make a statement, like, "Paul's got twenty dollars," some other kid would speak up and say, "He plays hob, too!"

Now, while I — or anybody else who had heard what had been said — knew exactly what the speaker had meant, that he didn't really believe that Paul had twenty dollars, I was still curious about what "playing hob" really meant. And to this day, I still don't know for sure. I've even done research on it and all I can find suggests that "playing hob" has something to do with playing the part of a mischievous person. And if that's the case, the way we used it back then makes absolutely no sense at all. Which does make sense, in a way, because we were always using nonsensical expressions when we communicated.

Take that Pat fellow I mentioned earlier.

Someone would say that so-and-so had gotten a job with the Northeast Coal Company and someone else would chime in with, "He won't last as long as Pat stayed in the army." And, just like "playing hob," no one could tell me who Pat was, what army he'd joined and how long he'd really stayed.

Research in that department revealed that Pat was not a real person but instead was the name the British used for a fictitious Irishman who was sort of incompetent. Apparently, then, if this was indeed the intent of the speaker in our case, he was meaning that the fellow who had been hired by Northeast was not too bright.

Even though the kids with whom I grew up weren't familiar with the word "slang," other than when we talked about our "slang shots," we sure were exposed to a lot of it. Sort of like somebody being as "ugly as home-made sin." And when somebody talked of being confused, he'd often state he "didn't know whether to wind his watch or go home." When someone who was one hundred percent sure of something would offer to bet a "shiny country nickel" on his knowledge, one had to wonder how much shinier a "shiny country nickel" was than a "shiny city nickel."

We'll probably never really know when and how all these expressions originated, but they sure are fun to ponder.

Having been born at the beginning of the last century and having spent his entire life in either Martin or Johnson counties, my father had speech patterns typical of most Eastern Kentuckians: Drop the "g" and go heavy on the double negatives.

"Gettin' pretty close to hog-killin' time," Dad would say. "Ain't no use gettin' in a hurry, though."

But as he spoke of that late fall/early winter ritual, those to whom he was addressing found absolutely no fault with his grammar. They were far more interested in the essence of the sentence he spoke than in its construction.

As a matter of fact, it wasn't until the middle of the last century that anybody else seemed to care, either. That's when some linguists out in a California university decided that Appalachians, to a man, had some sort of dreadful speech defect and sat about analyzing it, with the hope, no doubt, of "learnin' us better."

But alas, their efforts were stymied to a degree when one of their own decided that perhaps we were saying it right after all; that we were more true to the mother tongue than those who spoke it elsewhere. One thing that prompted that conclusion was a study of their own making that showed Appalachian high school students were much quicker to grasp the true meaning of William Shakespeare's writing than were those who ... well, didn't drop their "g's" and use a lot of double negatives. In other words, our kids made sense out of "What Ho, Banquo?"

As an English teacher, of course, I had to insist that the writing of my students be as grammatically correct as possible, and judging from the number of doctors, lawyers, and school teachers that passed through my classroom, I think I can say in all honesty that I had a bit of success in that department.

But we're not speaking of the written word here. The problem many "outsiders" have is how we sound. A friend told me once that her family had moved to a northern city for a short time when she was in grade school and that she cried all night after one of her new friends made fun of her after she had used the phrase, "Pon my honor."

Our thinking is that perhaps some northerners might could use a little learnin', too.

***

But as I've stated time and again, coal-camp kids never gave a thought to such, especially since so many really important things were happening. Like for instance pie socials — which were frequent fund-raising events at the H. S. Howes Community School.

In Loretta Lynn's movie, *Coal Miner's Daughter*, pie suppers were depicted exactly as I remember them. Boys old enough to be gainfully employed would bid on a pie and whoever bought it got to share it with the girl who had made it. Sometimes, depending upon how pretty the girl was — and coal-camp girls were a good-looking bunch — a pie would go for as high as six or seven dollars. The money paid for the pie went to the organization sponsoring the event. Most of the time, at least in our case, it was the PTA.

After the auctioneer — I remember that once it was Charlie Bailey, the preacher at the Free Will Baptist Church — would sell half a dozen, or so, pies, he'd take a break and we'd have us a cakewalk. Sometimes as many as eight or ten of the older women would each bring a home-made-from-scratch cake. Apple stack cakes were popular and there were always a couple of those up for grabs.

Those of us too young to have jobs and without enough money to bid on a pie would save what little we had for the cakewalks, especially since you could take the cake home and didn't have to eat it right on the spot with the woman who made it.

Clyde Roy Pack

With a piece of chalk, Mr. Garfield Chandler, the school's principal, would draw a line on the oily floor of the school auditorium and participants — at a dime a whack — would line up behind that line, sometimes 30 or 40 at a time. Someone would yell "Go!" and everybody would start walking around the room. After a designated time, whoever yelled "Go!" would yell "Stop!" and the person closest to the chalk line would win the cake.

Although I never won a cake, I enjoyed the competition. The same was true when we'd have dress-up day at school to celebrate Halloween.

We didn't wear store-bought masks. A mask, to us, was just a small strip of black cloth with eye holes ... like Zorro wore in the movies and the Lone Ranger would later wear on TV. What we wore — when we could get one — was a false face.

I haven't seen a false face like that in about 50 years, and as far as I'm concerned, it's good riddance. They were made of cloth and molded into what resembled a human face, complete with eye holes. They were made stiff with some kind of starch-like, gluey substance, but softened and lost their shape when they got wet. They were painted to resemble whatever character they were supposed to be and if you tried to eat through the tiny holes that had been left for mouths, you ended up eating the bitter, foul-tasting glue-and-paint combination. To top it all off they were held over your face with a thin rubber band that always broke the first time it was used.

As far as store-bought costumes, they were flimsy one-piece things that tied in the back like a hospital gown, and silk-screened to resemble Superman's suit or Frankenstein's monster. All in all, compared to today's elaborate Halloween fashions, they were sort of pitiful, to say the least. But we didn't pay much attention to stuff like that. Not more than one in ten kids in our neighborhood wore store-bought costumes that their moms had ordered from Montgomery Ward. The rest of us would just pull an old silk stocking over our faces, which seriously distorted our features, and we'd hide our regular clothes by putting our coats on backward, or something. Those of us who didn't use stockings would draw designs on our faces with lipstick or rouge or soot, anything to help disguise us.

I don't know who judged such events, but it seems that every year the boy who won the best costume had dressed up like a girl and the girl who won had dressed up like a boy.

Chapter 4

Neighbors in the News

Occasionally, a news item would appear in the newspaper giving ordinary folk the opportunity to see their names in print. This chapter is devoted to such items. Perhaps reprinting them here will resurrect long-buried bits of information, as well as preserve the memory of those who came before us and made our community special.

***

The Paintsville Herald, May 11, 1916 —*We have a Sunday school at this place of which we are all proud. The attendance is large and the interest manifested is great. The Sunday school and church work is under the supervision of Rev. T. J. Collins, who has labored hard and earnestly for the past three years to build it up. Bro. Collins is greatly assisted in this work by Mr. Glen Preston, who is also an earnest and loyal worker and has been a man of untiring energy in the church and Sunday school work here. Glen is the choir leader. He is an active, energetic, consecrated Christian, truly devoted to the cause of God and is beloved by all in this community.*

The death of Miss Lizzie Sparks, who is the daughter of Nelson Sparks, occurred at the home of Wm. Nickels Sunday morning. We mourn her loss. She leaves a father, three brothers and three sisters besides a host of friends. She also leaves a blessed memory and those who knew her best have no uneasiness with reference to her present state of eternal blessedness. The funeral service was conducted by Rev. T. J. Collins.

•

Bro. Collins was called Sunday to preach the funeral of Mrs. Fanny Welch, the widow of the late Thos. J. Welch. She died at the home of her son John, after a short illness resulting from injuries received by a fall. She was eighty years old and was married sixty years ago. Her body was carried to the cemetery near Concord by her grandsons.

•

Harry Howes says that owing to the advance of gasoline and lubricating oils, he will be forced reluctantly to dispose of his Ford car, and he now offers it for sale. Anyone wishing to purchase a Ford will do well to see Harry as he will sell at a sacrifice.

Clyde Roy Pack

***

The Paintsville Herald, October 5, 1916 — At 12 o'clock Saturday, Sept. 16, the death angel visited the home of Mr. and Mrs. Fred Whalen and took away their only child Carrol Kathryn, age 16 months.

She was taken very seriously ill on Monday with brain fever and during her short illness everything that loving hands and the best of medical aid could do was done for her, but God in His great wisdom thought it best to take her from this earth and plant her in heaven.

Little Carrol Kathryn was a very bright child for her age and will be greatly missed.

The funeral services were conducted by the Rev. T. J. Collins at Thealka church Sunday and interment at the Paintsville cemetery. Mrs. Whalen was formerly Miss Mollie Burton, daughter of Mr. and Mrs. E. L. Burton who are very well known in this community.

The bereaved family have the heartfelt sympathy of the whole community.

***

Coal-Camp Chronicles

The Paintsville Herald, March 15, 1917 — Mr. and Mrs. Glen F. Preston and Archie Fitch visited friends at Auxier Saturday and Sunday.

McClelland Preston spent Sunday with home folks at Mingo.

W. R. Davis was a business visitor at Auxier Tuesday.

Sterling C. Rice spent the weekend with home folks at Denver.

Miss Beatrice Salyer of Potter, is visiting her sister Mrs. J. N. Meek.

Mr. and Mrs. R. C. Burton were visiting at E. L. Burton's Sunday.

Misses Olga Stapleton, Vertrice Price, Lucy Rice and Lillian Thomas of Paintsville, were visiting Mr. and Mrs. Ray Stafford Sunday.

Clifford Rucker makes a daily trip to the post office this week.

Sterling Rice was calling on Miss Christina Burton Sunday evening.

We are having an excellent Sunday school at Thealka this spring with Bro. T. J. Collins as Supt. We have one of the best corps of teachers in the county consisting of Glen Preston, H. C. Howes, Mrs. Vess Hunter, Mrs. H. C. Howes and Mrs. Glen Preston, secretary.

Bro. Collins preaches every Sunday night with prayer meeting on Wednesday night. Everyone is invited to attend these services.

Messrs. Bert VanHoose and Hubert VanHoose have resumed their schools after an interruption of a few weeks on account of measles.

Thealka is a model mining town and has the best order of any similar place on the river, thanks to H. C. Howes, Assistant Supt., and the work of our Sunday school and church.

***

The Paintsville Herald, May 24, 1918 *— The week for the Red Cross drive closed with the name of every employee of the Northeast Coal Co. signed to the pledge cards agreeing to give the benefits of one day's work to the Red Cross. The total amount secured by the drive here was $1,601.09, and as the company agreed to match each dollar given by its employees, the grand total for Thealka will be $3,202.18.*

Sunday, June 16, is going to be one of the Red Letter days for Thealka, as we are going to have one of the greatest patriotic rallies as was ever held in Johnson County. We are going to raise one of the largest flags that has ever been flown in Johnson County. Arrangements are being made to have some of

Coal-Camp Chronicles

the best speakers in this end of the state to address the people. Music will be furnished by a band and local singing. Let everyone remember that date and come so that we can become better acquainted with present conditions and get the spirit of the times instilled in us as we never had before, then the Kaiser and his cohorts will begin to see that we are in this war to stay and that we will fight our way to victory and peace.

May our flag ever wave over the land of the free and the home of the brave — a haven of peace and rest for all nations and a symbol of right.

Messrs. W. R. Davis, R. C. Burton, McClellan Preston, Sterling Rice and Elmer Dawson, attended the flag raising at Auxier Sunday.

Miss Lottie Leek, who has been visiting her sister, Mrs. John Clatworthy, has returned to her home at Seco.

Rufus Maynard and Flem Conley have each purchased a new Ford car from the Big Sandy Hardware Co.

Sterling C. Rice was visiting in Ashland and other down-the-river points Saturday.

Misses Carroll Yeager, Margaret Dawson and Messrs. Virgil Picklesimer, McClellan Preston enjoyed a nice game of tennis Sunday evening.

Several from here were in Paintsville Monday night to see the boys who went to the army from Johnson County and to bid them God speed.

The Paintsville Herald, June 6, 1918 — *On Wednesday night the friends and relatives of Tabor Ward were shocked to learn of his death at Jenkins, which came unexpected to his many friends. His remains, accompanied by his wife and children, and Mr. and Mrs. Abray Bellomy of Burdine, and Alf Deboard of Keyser, arrived here Thursday evening and was taken to the home of Mr. & Mrs. M. G. Rister for the night.*

On Friday morning his remains were carried to the church house where a large congregation of his friends and relatives were gathered together to pay their last respects to the dead and in sympathy with his wife and three children. The funeral services were conducted by Rev. T. J. Collins of this place and he delivered a fine sermon from 1 Thessalonians 4:14, "For if we believe that Jesus died and rose again, even so then also which sleep in Jesus will God bring with Him."

Afterwards he was taken to Paintsville, and his body was laid to rest in the Preston Cemetery to await the last great call when his Master shall say, "Come unto my Father and inherit the place prepared for you, where death does not enter, nor moth and rust corrupt, neither do thieves break through and steal. A city prepared by God for his people, that needeth not the light of the sun and where there is no night."

His bereaved wife and family have the sympathy of the entire community.

Rev. Burns Conley of Paintsville preached to a large congregation at this place Sunday night.

W. H. Hughes and Millard Fraley were business visitors at Offutt Saturday where they went to employ teachers for our schools the coming year. They obtained the services of Mrs. F. J. Conley, Miss Fanny Weddington of Denver, and Mr. Charles VanHoose, and with this fine corps of teachers the prospects of good schools here is assured.

McClellan Preston, assistant Supt. Northeast Coal Co., was visiting home folks Sunday.

Thealka Sunday School is going to have a picnic June 16 and all the surrounding schools are invited to come and spend the day. Our Supt. Bro. Collins is making the arrangements for a good program.

In the afternoon the patriotic citizens of Thealka are going to raise a large flag and the crowd will be addressed by some of the county's best speakers.

Music furnished by a band and local singing. Let everyone come and enjoy the day.

***

The Paintsville Herald, June 20, 1918 — *Last Sunday one of the largest gatherings in the history of Thealka assembled to raise the large American flag the company and the men had bought for the occasion. The crowd gathered in the forenoon and at 12:30 the flag was raised by Misses Carroll Yeager and Margarette Dawson while the Paintsville band played the "Star- Spangled Banner."*

Raising the American flag on June 16, 1918.

After the flag was raised the crowd assembled on the hill where speeches were made by Hon. Jas. W. Turner, Manager H. LaViers and Judge Fred A. Vaughan.

It was a great day for Thealka and the program was carried out in an interesting and pleasing manner. Sterling C. Rice, a bookkeeper for the company, had the matter in charge and all who attended were pleased with the entertainment. Dinner was served on the ground

***

The Paintsville Herald, October 9, 1924 — There was a large crowd attending church at this place Sunday night.

Mr. and Mrs. Charley McKenzie were the Saturday night guests of friends and relatives at Sitka.

Misses Elizabeth, Ruie and Rice Castle were the all-day guests of relatives in Greentown Sunday.

Mr. and Mrs. Proctor Robinson and family were the all-day guests of Mr. and Mrs. W. M. Robinson.

Miss Myrtle Hamilton was the Sunday night guest of Misses Emaline and Adaline Robinson.

Born to Mr. and Mrs. Luther Sturgill, a fine baby boy.

There will be a prayer meeting at Boyd Branch Thursday night.

Clyde Roy Pack

***

The Paintsville Herald, December 18, 1924 — *The Northeast Coal Co. mines have been working straight time for about three months. We understand they have not an empty house in the camp.*

The $20,000 high school building here is going to put the finishing touch to Thealka. It will be completed by January 1, 1925. This school will be an inducement to hold good miners and to cause other good miners to come to Thealka in order that their children may obtain high school training at less cost than to send them to other towns or cities.

Mrs. Lou Compton of Catlettsburg, was visiting her husband Saturday and Sunday. Mrs. Compton was Miss Eleanor Ferguson before her marriage a few months ago. She was formerly of this place.

Hobert Hayes was calling on his aunt, Mrs. Flem Griffith Sunday night.

Misses Claudia and Elizabeth Butler of Paintsville were visiting Miss Guthrie Louise Griffith Saturday night.

The box supper at Boyd Branch school house Saturday night was given in order to help with the Christmas entertainment. There were not many girls and boys out.

Coal-Camp Chronicles

Uncle Frank Burgess passed away at his home on Dec. 12. Funeral services were conducted by Rev. J. S. Hager and others. The body was laid to rest in the S. P. King Cemetery, Dec. 13. Uncle Frank was a good man and the whole community will miss him.

Miss Roberta Thurman of Huntington, W. Va., attended the funeral of her grandfather, Frank Burgess.

Joe Castle, who happened with an accident on the Paintsville railway yards when an engine hit him, is very much improved and will soon be out among the boys again. Mr. Castle is one of the best civil officers of the state.

***

The Paintsville Herald, January 15, 1925 — Misses Emaline and Adaline Robinson gave a birthday party Wednesday night Jan. 7. Those present being Alma Castle, Ruie Castle, Elizabeth Castle, Mabel Castle, Mildred Conley, Rusha Conley, Dixie Rice, Vivian McCloud, Hazel Spears, Sophia Bush, Minnie and Ethyl McKenzie, Ivel and Gailord Castle, Stinson Conley, Junior Wyatt, Eugene Davidson, Solon Conley, Tommy McFaddin, Shady Preston, Nathan Rice, Isom Lewis, Charlie Bailey, Tracy Pack, Tive Daniel, Frank Daniel, Ruel Gibbs.

Many presents were received and games played. At a late hour the party broke up, all saying they had had a nice time.

Mr. and Mrs. Russell Conley who have been on the sick list are improving.

Mrs. Christina Stapleton of Detroit, Mich., is here spending a few weeks with her father and mother.

Miss Hazel Spears was the weekend guest of her uncle at this place.

The stork visited the home of Mr. and Mrs. Flem Osborne and left a four-pound baby.

John and William Robinson were the weekend guests of their uncle Sam Spears of Lick Fork.

Mr. and Mrs. Charley McKenzie were the Saturday night guests of his mother, Mrs. Whitten.

Miss Adaline Robinson was the Friday evening guest of Mrs. Rosy Spradlin.

Miss Ella and Emma Spears were the Saturday night guests of Mrs. Rosy Spradlin.

Walter Rice of Jennies Creek was the guest of his sister, Mrs. Grant Johnson, Wednesday night.

Miss Ruby Wyatt, Mrs. Willard Davis and Mrs. Christina Stapleton were calling on Mrs. Jim Preston Wednesday night.

We are glad to announce that the little infant of Mr. and Mrs. Roy Colvin who has been very ill with pneumonia is able to be out again.

Miss Sophia McCloud who is teaching school on Rock House is at home visiting home folks this week.

***

The Paintsville Herald, June 11, 1925 — Thealka defeated Paintsville (6-3) at Paintsville last Saturday afternoon in a good game. It was one of the hardest-fought games of the season in the new league. Many of the fans were surprised at the high-class ball playing of the Thealka team. This team is composed of young players but they have been working as shown by their playing.

Dan Pugh pitched big-league ball for Thealka and was never in danger. His team got a big lead in the fifth inning and held it until the close of the game. Paintsville made three scores in the seventh and it looked like they were going to tie the score. Pugh retired the first three men up in the last half of the ninth inning and struck out ten men in the game.

Wells pitched the first seven innings for Paintsville and pitched good ball, but the heavy hitters of the Thealka team connected with his balls for five runs. Ward relieved Wells and pitched the last two innings, striking out two men.

It was a nice game. Miller, the 14-year-old shortstop for the Thealka team, played star ball. Puckett, the manager of the team, was right there all the time.

Players for Thealka included Preston, Puckett, VanHoose, Colvin, Pugh, Conley, Miller, Colvin and Castle.

***

The Paintsville Herald, November 29, 1928 — Another candidate for the office of jailer of Johnson County is reported this week from the little city of Thealka in the person of Foster Burton, a citizen of that section who is one of the best-known men in all the county. Mr. Burton makes his announcement after being solicited by a large number of his friends who say he can win the nomination in a walk. He has never been a candidate, never held an office and is in every way qualified to fill the position to the entire satisfaction of the people of the county. He has given his promise that he will take special care of the county's property, and attend to all the duties of the office of jailer.

Coal-Camp Chronicles

He has been a blacksmith for more than 25 years and is a hardworking, honest, progressive citizen and would make the county an excellent official.

***

The Paintsville Herald, March 7, 1929 — *Misses Irene and Pauline Burton spent Friday, January 22, at White House visiting friends.*

Mrs. Arthur Mosley and children of Drift, Ky., have returned home after an extended visit with Mr. and Mrs. Edgar Preston.

Claude C. Preston, who underwent an operation for appendicitis at the Paintsville Hospital Tuesday, February 5, is recovering rapidly.

Mr. Cyrus Preston and son, Ernest Ray, spent Sunday, February 10, in Ashland, visiting Mr. and Mrs. P. D. Spears.

Mr. Eugene Dawson, of Ashland, spent Friday, February 15, at this place visiting his sisters, Mrs. Sherman Fitch and Mrs. Virgil Green.

Miss Mona Daniel spent the weekend in Ashland visiting Miss Irene Adkins.

Miss Kathleen Stapleton spent the weekend visiting her parents, Mr. and Mrs. Beecher Stapleton, of Manila, Ky.

Charlie Staggs returned from Ashland Thursday, February 21, where he had been called by the illness of his son, Clyde.

Mrs. Sonnie Litteral, who has been a patient at the Golden Rule Hospital for a few weeks, has returned home.

Born to Mr. and Mrs. Alfred Sherman Friday, January 22, a fine baby girl, named Fannie.

Mr. and Mrs. Stelson Conley announce the birth of a son Saturday, February 2, named Charles Odes. Mrs. Conley will be remembered as Miss Loula Childers.

Mr. and Mrs. Estill Travis announce the arrival of a daughter, Tuesday, February 19. She has been given the pretty name of Louisa.

Mr. and Mrs. Monroe Castle are the proud parents of a son which registered at their home Monday, February 18. He has been named Billy Ray.

***

The Paintsville Herald, June 23, 1932 — *The post office at Thealka was entered by robbers last Sunday night, but no valuables were missing. The mail was gone through by the thieves and the office disarranged and office supplies scattered about, but it is not thought that any mail was taken. The robbers* gained entrance to the building by ripping off the window screens and breaking the window panes.

An investigation is being made and it is believed that the guilty parties will be apprehended. Finger prints of the robbers were taken and arrests are expected.

Mrs. Lizzie Colvin is postmistress at Thealka.

Quite a number of attempted robberies of post offices have taken place in Johnson County in recent months.

***

The Paintsville Herald, March 26, 1942 — *Those who visited Mr. and Mrs. Frank Castle Sunday were Mr. and Mrs. John Dills of Nippa, and Lieutenant Harry Huff of Fort Knox.*

Miss Nadine Lyons was the Thursday night guest of Misses Mildred and Sylvia Castle.

Alfred Crider was the Sunday guest of his daughter, Mrs. Woodrow Castle.

Mr. Andrew Castle attended the show at Paintsville Saturday night.

Mr. and Mrs. E. B. Jackson were the Saturday night guests of Mr. and Mrs. Woodrow Castle.

Mr. and Mrs. Beecher Castle are the proud parents of a fine baby boy.

Mr. and Mrs. Willie Pack are the proud parents of a fine baby boy. He has been named Willie Joe. (Yep! This is little brother Joe who became my buddy and chief playmate for many years. He was Tonto to my Lone Ranger).

A large crowd from Tutor Key attended the revival meeting here Saturday night.

Sylvia and Mildred Castle, Lieutenant Harry Huff and Marshall Meade attended church at Tutor Key.

Mrs. Tom Fitch is very ill at this writing. We hope for her a speedy recovery.

Mr. and Mrs. James Lyons and son Ernest Keith, were the Saturday night guests of Mr. and Mrs. Ernest Dove.

Woodrow Castle and daughter, Jessie, made a business trip to Paintsville Saturday.

The revival at Thealka is progressing nicely.

Misses Clara Church and Fannie Castle of Tutor Key were the Sunday afternoon guests of Miss Mildred Castle.

***

The Paintsville Herald, March 18, 1943 — *Mr. & Mrs. Herbert Castle were called to the bedside of Mr. Castle's mother who passed away Saturday night.*

Those attending the farewell party given in honor of Pvt. Don Franklin were Marie McKenzie, Goldia Ray Preston, Mildred Castle, Betty Helen Conley, Mr. Douglas Castle, Mitchell Lyons, Carl McKenzie and Paul Preston.

Miss Sylvia Castle and Mildred Castle were visiting Mrs. Emma Meade and Mrs. Tollie Meade Sunday evening.

Mr. & Mrs. Woodrow Castle were visiting Mr. and Mrs. Alfred Crider Saturday night.

***

The Paintsville Herald, April 1, 1943 *— Mrs. Luther Castle has received a letter from a Mrs. Alexander, of Dundee, Scotland, in which the Scottish mother writes of her acquaintance with Mrs. Castle's son, Gene. The letter follows:*

Dear Mrs. Castle:

You will be surprised to hear from me, a Scottish mother who has a son away in the forces just like yourself. I am Mrs. Alexander whom I understand you have heard about from Gene. He was stationed in our little town here and I saw quite a lot of him. I kept two of the boys of his company who were Gene's pals and Gene stayed quite near to our home. I hope you are receiving mail regularly from Gene as that is about all we can do these days, although from what we hear from the wireless they are getting along alright in Africa. Gene told me about his young sister and brother. I am sure they must miss him very much. I have only two of a family, one boy and one girl and when my son joined the Royal Air Force we missed him very much, but we have just to get used to these things in war time. My boy is stationed a long way from home but we are thankful that he is still in this country. Now Mrs. Castle, I hope that you don't mind me writing to you, with kindest regards from all.

Respectfully,

(Mrs.) Alexander

***

The Paintsville Herald, June 3, 1943 — *Fort Dix, N.J.: Helen I. Dale, of Thealka, Ky., who enlisted in January in the Women's Army Auxiliary Corps, is now on active service as a supply clerk at Fort Dix with the 42nd WAAC Headquarters Co. Auxiliary Dale is the daughter of Mr. and Mrs. James Dale. She received her WAAC training in Florida this winter at the Daytona Beach Training Center, and reported for duty with her company at Fort Dix several weeks ago. The WAACs at Fort Dix have replaced soldiers in twelve different non-combatant assignments, releasing the men for combat service.*

***

The Paintsville Herald, April 20, 1944 — Mrs. Ruth Castle, Thealka, has received a telegram from the War Department that her son, Staff Sergeant Gene Castle, is being held by Germany as a prisoner of war.

***

The Paintsville Herald, December 14, 1944 — The War Department has notified Mrs. Minnie Castle, Thealka, that her son, Pvt. Sam B. Castle Jr., has been slightly wounded in action in Germany. He is 19 and has been overseas three months, serving in Italy, France and now in Germany.

His three brothers in service, all with the Army overseas, are S-Sgt. Ernest Castle, in France; S-Sgt. Curtis Castle, in Dutch East Indies; and Pvt. Bruce Castle, who has just recently embarked for overseas duty.

***

The following item was of special interest to me. It marked the first time that I had lost someone near my age to death. I wrote about it in my book, *Muddy Branch, Memories of an Eastern Kentucky Coal Camp*.

A portion of that segment follows:

... a group of us had played baseball until it had gotten so dark we could hardly see each other, let alone the ball. Sweaty

and tired we had walked home amid the "catch ya later" and "see ya tomorrow" farewells, as one by one each player would come to his house.

The next morning, the first thing Mom said to me was, "Eck's dead."

To me that was impossible. Just yesterday we had played ball together. He was fine. He couldn't be dead.

But he was.

His appendix had ruptured during the night and he died before they could get him to the hospital.

Eck's death literally devastated me. I had seen lots of dead people: old people or not-so-old people who had been killed in the war or in the mines. But not someone just like me. I remember that they ran a school bus from Meade School so that Eck's fellow students could attend his funeral at the Thealka Free Will Baptist Church.

***

The Paintsville Herald, August 24, 1950 — Eskel Lee Castle Sparks, 15, died in the Paintsville Hospital at 10:30 a.m., August 19, as the result of a ruptured appendix on the same day.

Born at Thealka on March 31, 1935 and a resident there all his life, he was the foster son of Mr. and Mrs. Jeff Sparks.

Surviving besides his foster parents are three brothers and two sisters: Esteel B. (twin brother), Joe, Junior, Elizabeth Mae and Darlene Castle, all of Barnetts Creek.

Also surviving are four foster brothers and sisters: Tom, Ray, Virginia and Faye Sparks.

Funeral rites were held at the Thealka Church House at 10:00 a.m., August 21, with the Revs. Cully Sparks and Florda Lyons officiating.

Burial was made in the family cemetery at Thealka, under the direction of the Preston Funeral Home.

***

Although I knew Eck even before I ever started school and we more or less grew up together in Society Row, I must have been eight or ten before I learned that he wasn't Jeff and Doll Sparks' real son. I doubt I really knew what it all meant, but I suppose he was the first foster child I ever knew.

And it was even longer, and quite by accident, that I discovered he had an identical twin.

That in itself is another story.

Anyway, one day when I was in my back yard doing nothing in particular, Eck came running down the back lane.

"Hey, Eck," I yelled.

He threw up his hand and went on by.

As I turned to go toward the front of the house, not more than 15 seconds after I'd seen him running down the back lane, Eck ran by again, this time going in the opposite direction, almost as if he were chasing himself.

It's been more than half a century ago, but I can imagine the first thought that went through my head was that Eck sure was a fast runner.

Of course, when I ran to Mom and told her what I had just seen, she explained all about his being a twin and everything.

Eck Sparks

Chapter 5
Around the House

As a journalist, I've had the opportunity over the past few years to interview and write the thoughts of dozens of Eastern Kentucky's older citizens. Consequently, I think I've gained a degree of personal insight as to just how much of a treasure these older people are.

I've found that whether I'm talking to a 90-year-old former coal miner, a great-great-grandmother who has never been more that 50 miles from her birthplace and never worked a day in her life outside her home, or, a life-long fox hunter, what I'm hearing are genuinely honest slices of life.

Ironically, while many times the views of the older citizenry are in direct contradiction to what so-called scholars have recorded in history books, I find myself knowing, without doubt, that I'm listening to the true story of how it really was. The cold, slick pages of some thick anthology can't possibly explain the humanity involved in trying to feed and clothe a large family during the Depression, or make you feel the heartbreak of shipping sons and daughters off to war. As the voices of those who've lived the events of which they speak sometimes break and become almost a whisper, history itself becomes a living, breathing entity; a face with eyes that sometimes smile, sometimes cry; a history with a soul.

After becoming an adult I've wished dozens of times that I'd had sense enough to encourage my own parents — Willie and Julia Baldridge Pack — to talk about their pasts. Even though I doubt I could ever have gotten them to actually write down their thoughts, as I look back I can now see how simple it would have been for me to tape them.

Willie and Julia Baldridge Pack

I know absolutely nothing of how they felt about anything when they were young. I know nothing of their courtship; how they coped during the early years of their marriage; even how

Coal-Camp Chronicles

they arrived at their religious and political preferences. (Free Will Baptist and Democrat.)

Admittedly, knowing all this may not have made an iota of difference as to how I, or any of my six siblings, turned out. But as I get older, I find myself wondering about Dad's and Mom's philosophies; about family traits that are exclusively ours; about any little quirks that set our family apart from all the other families that inhabited Muddy Branch during the 1940s.

I sorely regret not encouraging my folks to write, or at least dictate, their memoirs.

***

My father, Willie Pack (1900-1969), was probably as good an example of an Eastern Kentucky coal miner as one could ever find. Although many writers in the past, for reasons that are beyond me, have depicted coal miners as hard-drinking, hard-fisted rabble-rousers, Dad was more like ole Wild Bill Elliott in the Saturday matinees I'd watch on a regular basis at the Sipp and Royal theaters in Paintsville. He was a "peaceable man."

In the 30 years I knew him, I was never aware of his quarrelling with a neighbor, or even appearing to be out of sorts with any of them. Even after he discovered the identity of the thief (a fellow miner) who — in the dead of night — carried off a whole ham that Dad had curing in the smoke house not 20 yards from our back door, he confronted him peacefully, merely asking him if he had a ham sandwich in his dinner bucket as they ate lunch together in the corner of some dark mine shaft deep in the Northeast Coal Company's Number Three mine.

Mom said later that the man had a big family of kids and Dad would have given him the ham had he asked for it.

My father was indeed a generous man, known throughout the community for his benevolence. On many occasions I was drafted to tote vegetables he'd raised (he always tended a big garden in the head of Slaughter Pen Hollow) to older neighbors and widows living nearby. Of course, our pantry never suffered because he grew more than enough of everything and had he not given some of it away, it would have rotted on the vine, or in the ground, as the case may have been.

Although he had been poorly educated, he read the Bible every chance he got. The complicated sentence structure and subtle references of the King James version never deterred him as he'd read and re-read a chapter until he felt he had grasped its meaning. I think being a deacon at the Thealka Free Will Baptist Church, he felt it was his duty to "know as much Bible" as he could know. Although I'm certain he never once challenged any of the preachers on their interpretation of certain Scriptures, I have heard him tell Mom as we'd walk home from church on a Sunday night that the preacher in question "didn't get it right." Knowing how often Dad read the Scriptures, I never once doubted what he was saying .

Dad was also very much against church politics and was often critical — to only Mom, of course — about some decision or the other made at the monthly business meetings.

Although physically hardened from more than 40 years as a coal miner, and although his back was bent and his hands calloused, he was one of the most gentle men I've ever known. He never seemed to question his lot in life and took pride in the fact that the coal he dug from deep within the earth not only put food on his family's table, but also kept the country on the move. He also took great pride in the fact that he had always provided a full day's work for the wages he'd earned.

But more important than anything else, he was always a peaceable man.

***

Long before I was old enough to go to school, I was constantly carrying things home I'd found in and around the neighborhood. It wasn't unusual for me to show up at the back door with a brand-new pup or kitten to introduce to Mom. Sometimes she'd let me keep my new acquisition, sometimes she wouldn't. Sometimes she'd simply say, "Let's see what Daddy says about it."

One day I brought something home that caused quite a stir at the Pack house. I've no idea where I picked it up, but I can remember thinking how pretty it was; how pleasant the new addition to my vocabulary sounded; how it just rolled right off my tongue; how I couldn't wait to share it with Mom and Dad and little brother Joe.

Well, I shared it all right, right at the supper table, somewhere between "pass the soup beans" and "I want another glass of milk." I guess all that saved me was the fact that it was pretty obvious that I had no idea that the phrase I had just so casually uttered was obscene. And not just regular cuss-word obscene, but downright dirty obscene. But before the echo of the words had faded from the walls of our little company-house kitchen, I was experiencing a ringing in my ears that proved perfect accompaniment to the lyrics being sung by both my parents.

In the 60 or so years since that moment, I've never used that phrase again.

Loretta Lynn sang a song once that said, "They don't make 'em like my daddy anymore." Well, mine either, Loretta. If they did, there'd be no ugly language permitted in his presence, especially from children.

***

As we sat on the deck on a recent summer night watching it get dark, Wilma Jean and I began talking about how the summer had gone so quickly, as had our youth; a time more than half a century ago when sitting outside watching it get dark was our primary form of entertainment.

It's been a lifetime ago, of course, but at times it seems that only yesterday television was a thing of the future and the radio battery had to be saved for special occasions, like the news and the Grand Ole Opry on Saturday nights.

Anyway, as conversations are apt to do, ours wandered aimlessly before our discussion turned to things we did as kids to entertain ourselves during daylight hours. I spoke of swimming in the Number One Pond and playing cowboys. But her memory provided the most entertaining story of the night. She remembered the time when, as a very little girl living in Williamsport — located in the far eastern part of Johnson County and the community that housed Meade Memorial High

School — that she and some friends, while taking a walk in the woods, discovered a moonshine still, complete with dozens of shiny clear fruit jars all lined up ready to be filled. Unfortunately, the still was unattended at the time (or perhaps the lawbreakers had headed for the bushes when they heard the small army of Two-Mile Creek kids approaching), so the moonshiner remained anonymous and all Wilma Jean got for her troubles were a few scratches around her pretty little ankles.

Guess my life was not as exciting as hers as there are no moonshiner stories in my past. Moon pies, yes. Moonshiners, no. Nevertheless, some things that occurred when I was a youngster are indeed worth remembering. For instance, when I was about seven years old and brother Joe was about five, we met in the kitchen on a sweltering summer day, both heading for the refrigerator.

Our objective was a delicious chilly imp. (For those not ice cream literate, a chilly imp was vanilla ice cream with a chocolate coating .. on a stick. Wilma Jean said they called them "sailor bars" on Two-Mile.) Ordinarily, such a treat never made it beyond the steps of the Northeast Coal Company store, as we'd plop ourselves down, gobble it down and lick the stick before we ever made a move toward the house, or wherever our next destination happened to be. But apparently, Dad had brought home a couple of extra this time. The problem was, as we opened the tiny freezer compartment of our Crosley Shelvador, we discovered that where there were supposed to have been two, there was only one chilly imp left. Either Joe himself or older brother Ernest had snitched the other one when I had been preoccupied with my chores. (Well, somebody had to read the latest issue of Blackhawk comics.)

Now, had our prey been a popcicle, our dilemma could have been easily solved, because popcicles, as everybody knows, had two sticks, were grooved down the middle and, with a little practice, were easily broken into two equal parts. Not so the chilly imp, however. It boiled down to all or nothing.

War was emminent. Brother against brother. The Civil War revisited.

As four hands wrapped around the frosty treat simultaneously, we both yelled, "Mine!"

Even though I was at least a head taller than my little brother, he was adamant and stood his ground. I really think he would have fought me over that chilly imp.

"You had the last one," he yelled.

"Did not! You did!"

"Did not!"

"Okay," I said, "I'll flip you for it."

We placed the chilly imp on the kitchen table and ran out into our grassless backyard where I hunted a little flat rock about the size of a half dollar. I spat on one side of it, rubbed it in real good and said, "Call it."

As I flipped it over our heads, Joe yelled, "Wet!"

The tiny rock landed between us at our feet, wet side up.

"It's wet! I win," Joe said as he turned to go back into the house.

"Wait," I said, "we gotta go two outta three."

Surprisingly, he agreed and as I flipped the rock again, he yelled, "Dry!"

It again landed at our feet ... dry side up.

"Dry," he said in a sort of "I told ya so" voice, and again started toward the kitchen door.

"Wait!" I yelled. "We gotta go three out of five."

Reluctantly, he again agreed, but again he called it correctly.

We walked back into the kitchen together and as he picked up the chilly imp, it fell off the stick onto the kitchen floor, a shapeless mass of brown and white goo that was immediately pounced upon by Tiger, our old one-eyed cat, who had followed us inside as the screen door slowly closed behind us.

As would be expected, Joe and I had lots of disagreements over a variety of things. Fortunately, none of them were ever serious.

***

Except for April 10, which is my birthday, and December 25, of course, the dates on the calendar meant little to me when I was a kid. But, despite the fact that I hardly knew what day of the week it was, I always knew when it was the Fourth of July because of the pop.

On weekdays Dad would stop at the company store on his way home from work and pick up staples, like bread or bologna or a can of carbide, those little gray pebbles that provided fuel for the little lamp that miners wore on the bills of their caps when they entered the darkness of the mines. But on Saturday, he and Mom would go to Roy Melvin's store in town and shop for groceries. Roy, or one of his delivery boys would deliver them in a pickup and set them down right on our front porch. That made things a lot more simple since Mom and Dad usually had to call Fat Blair and his taxi when they'd go to town. And, without fail, on the weekend before the Fourth of July, there'd be a case of pop —

all mixed up in various flavors. There'd be a bottle or two of Spur, RC or Pepsi, but the rest would be Nehi fruit-flavored, like strawberry, orange, grape and peach.

With only 24 bottles to a case, sometimes there would be only one bottle of a particular flavor. Therefore, as soon as the delivery man would leave, Joe and I would pounce on it like a hungry chicken would a june bug and stake our claims to our favorite colors.

"I get the red one," he'd say. "And one of the oranges."

Sometimes we'd argue over a particular bottle, many times settling the issue the same way we did when we argued over the chilly imp.

Of course, if Dad or Mom or one of our older brothers or sisters — unaware (and caring less) of the way Joe and I had divided up the bounty — beat us to one of our choices, we had no other choice but to settle for whatever was left.

I guess I was grown before I ever saw a real Fourth of July fireworks display. As a matter of fact, I was in college and at a carnival in Richmond, Kentucky, the first time I ever witnessed a real, honest-to-goodness fireworks display. I'm sure the Paintsville Fire Department put on a fireworks show on the Fourth, but I had never seen one.

But when I was a kid, for the special occasion Dad or brother Ernest would come up with a pack or two of firecrackers. There would be about 30 in a pack and Joe and I, being careful not to pull out a single fuse, would separate them. Much to the distress of neighborhood dogs, some of the kids would light a whole pack at a time. But not us. Firecrackers were too hard to come by so we enjoyed them one at a time. Even when one fizzled out and didn't go off, after making sure it was indeed a dud, we'd break it in half, dump out the powder and light it. It wouldn't make a bang, but would flame out like a sparkler and make a lot of smoke.

All this was preceded, of course, by the story Mom would tell us about the nameless little boy who a long, long time ago in a galaxy far, far away (I just threw that part in about the galaxy) wasn't careful and let a firecracker go off in his hand and it blew off his fingers. I'm not sure, but I think he must have been a brother of the other nameless little boy she'd always tell us about during dog days. It seems he went swimming one time with a chigger bite on his arm, took lockjaw, and after his arm rotted off, died. We never really doubted her stories, but at the same time, pretty much ignored her warnings.

Anyway, I doubt I'd even have known what they were talking about if someone had mentioned the Declaration of Independence. Had it not been for fruit-flavored pop and firecrackers, the Fourth of July would have pretty much been like most other summer days .

***

As I mentioned earlier, despite the fact that he was a full-time miner, Dad tended a big garden in the left-hand fork of Slaughter Pen Hollow. For some reason I was considered pretty much useless for any other garden-related chore, so I'd get volunteered to start packing pokes filled with surplus vegetables to the widows and elderly neighbors who lived in, or around, Society Row.

Pictures of Dad's gardens could have served as covers for the dozens of seed catalogs that lay among old copies of the *United Mine Workers Journal* and the latest copy of *Grit* that Mom kept in a neat little pile at the end of our couch. No weed in its right mind would dare grow between his corn rows, and the bounty of that little half-acre plot that he and older brother Ernest had fenced in with barbed wire to keep the wandering livestock honest, more than matched the garden's neatness.

Anyway, not only did Dad grow enough tomatoes, corn, potatoes and green beans to feed us, but he also spread, mostly via me, his benevolence from one end of the camp to the other. He obviously had a green thumb when it came to growing stuff and even after Mom would can dozens of jars of everything she could can, there seemed to be plenty left over.

He would have been insulted if a neighbor had offered to pay him for anything he'd grown, but when one of them would slip me a quarter now and then, that was just fine with me. After all, a man had to do what a man had to do to raise enough for show fare at the Saturday matinees. Not that I suffered in that department, what with Gene Miller paying 15 cents a case for pop bottles at the company store.

Although Dad's generosity was well known throughout the camp, especially as far as his vegetables were concerned, it did have its limits. For example, the head of one large family that included three unemployed young sons whose ages ranged from about 17 to 21, and whose work ethic matched the Tussie boys from Jesse Stuart's novel, was always hinting around to Dad for a good mess of his beans or a couple of his big tomatoes. But not once did Dad ever instruct me to deliver anything to them.

"Those boys won't strike a lick at a snake," he'd say, and obviously referring to 2 Thessalonians: 3:10, he'd add, "If a man will not work, he shall not eat."

Yet surprisingly, I did hear him tell the boys' father that he could have all the beans he wanted out of his garden, but those boys would have to pick them. As far as I know, though, they never did. I think Dad knew they wouldn't.

Of course, my father wasn't the only Society Row miner who raised a garden. By the same token, neither was I the only kid who was volunteered to tote fresh vegetables to some of the needy neighbors.

Sharing in that manner was simply typical of the times.

***

In 1928 a St. Louis pharmacist, Jim Howe, took his wife on an ocean cruise. Before leaving, he concocted a remedy for his wife's indigestion ... and invented Tums.

That tidbit of information is from a story I read in the newspaper in late 2003, which reported that Tums was indeed invented 75 years ago. I can't imagine why the age of such a product would merit a newspaper story, but even more of a mystery is why I'd bother to read it. Then again, since its birthday supersedes mine by ten years or so, the product is part of my childhood memories. You see, when my father would empty his pockets upon returning from a shift's work in the Northeast Coal Company's Number Three Mine, a partial roll of Tums very often accompanied his old Barlow knife, a small tangle of tiny multi-colored wire, and a mostly-used-up roll of (what we called) miner's tape.

I knew that Tums aided indigestion and if I gave it any thought at all, it was probably that Dad was the only miner anywhere who carried a roll with him to work. However, after I became an adult, I came to realize that a roll of Tums was pretty much a staple for coal miners.

But when I was ten years old, I knew nothing about anything like that and as far as I was concerned, the chalky, mint-flavored tablet was just another piece of candy to be gobbled down when Dad or Mom wasn't looking.

By the same token, the aforementioned article, tucked away in some obscure section of the newspaper, has simply provided still another memory of a time more than half a century ago when the contents of a father's pockets were a curiosity to a small boy.

Chapter Six

Learnin'

When I started school in the fall of 1945, I was privileged to attend one of the finest grade schools in the county. As a matter of fact, the H. S. Howes Community School (named for the Paintsville attorney who provided the land upon which it was built) was likely judged at the time as being state-of-the-art. There were probably 50 or 60 schools in the county in those days, many of which had one room. But the big yellow building where I began my formal education had five classrooms (which means there were five teachers, and sometimes a principal who didn't teach any classes) and a huge auditorium. Well, it seemed huge to me when it filled once a week for the singing teachers. Then again, it wasn't big enough when I'd stand in line to face the health nurses who came once a year to give us our shots and check us for head lice.

The building also had indoor plumbing, which was really special to us at the time since none of the houses in the community were so equipped, at least none of which I was aware. I'm not sure, but I think the company store could have had an indoor bathroom.

To me, the school building always provided a feeling of comfort, safety and warmth, and as an early elementary student, next to my own house, of course, it was the place I most liked to be.

As nice as it was, however, it wasn't the building that mattered, but the teachers. In the eight years I attended school there, more than two dozen dedicated individuals, who no doubt worked for meager wages, stood before eager coal-camp kids and drilled us in the three R's. Whether or not we realized it at the time, they were likely also giving us life skills that enabled us to compete in what would eventually become an extremely complicated world.

The only test we had, except those the teachers gave us in the various classes, was what we called (and dreaded) the "eighth-grade exam." Everybody always passed, but the rumor was that if one failed, he couldn't go on to high school. Most of us Muddy Branch kids went to high school at Meade Memorial. Some went to Paintsville. Once in a while, some didn't go ... period.

I'd love to see one of those eighth-grade exams we took back in 1953 and compare it to some of the tests kids take today. We had to do math without calculators and really had to know how to spell a basic set of words. We really had to know that it was *i* before *e*, except after *c,* since word processors and spell check were half a century away.

Oh, and one more thing: every teacher had a paddle ... and knew how to use it.

Having grown up in a coal camp, I suppose it's only understandable that I lacked a certain degree of sophistication. After all, although both my parents were wonderful people — the salt of the earth, really — they too had been reared rural poor in the early 1900s.

Consequently, Mom had only a grade-school education, and due to tragic circumstances that left him the bread winner for a houseful of siblings when he was in his early, early teens, Dad didn't even have that. So, the only magazines read at our house were *The Progressive Farmer* and the *United Mine Workers Journal*. As I mentioned earlier, we depended upon the radio and, once we got TV, John Cameron Swayze and his Camel News Caravan (Camel cigarettes sponsored the show), for our national news. Before I enrolled at Eastern Kentucky State College in Richmond in the fall of 1957, about the only newspapers I ever read on a regular basis were *The Paintsville Herald* and the *Grit*, the weekly out of Williamsport, Pennsylvania (now it's published in Kansas). But when I moved into Memorial Hall (the oldest mens' dorm on campus at the time), from a flier someone had slipped beneath my door, I learned that the Louisville *Courier-Journal* was running a special for college students. I can't remember exactly the terms of the deal, but I took advantage of it and for pennies a day, found the *C-J* at my door every morning. That began my life-long love for waking up to the morning paper. These days, it's the *Lexington Herald-Leader* that provides the cream for my coffee and helps me jump-start my day.

So, even before I dressed for class, I enjoyed searching for "Lois" in Hugh Hayne's editorial cartoons (For non-Hayne fans, Lois was the celebrated cartoonist's wife and he would hide her name, sometimes as many as half a dozen times, in his drawings).

But perhaps even more important than searching for Lois, I became a regular reader of columnist Allan Trout. I liked his stuff for two reasons: He was a quick read and, more often than not, provided a bit of humor. (As my grades will reflect, I was apparently really big into humor in those days.)

Anyway, Allan Trout came to mind again all these years later when a friend gave me a couple of Trout's collected works. Although I've yet to read anything that strikes a familiar chord

— not that I'd expect myself to remember anything I'd read that many years ago — this collection apparently consisted of columns he'd written and published at about the time I discovered him.

But when I began reading Trout as a pimply-faced college freshman almost half a century ago, I had no way of knowing that someday I would be charged with producing 52 columns a year for *The Paintsville Herald* and a couple of other newspapers in Eastern Kentucky. And even if I had, I sure would never have been so egotistical as to think that his work and mine would have anything in common. However, after reading an entry in one of my newly acquired volumes, I've decided that is indeed the case.

In this particular article, Trout was telling a delightful tale about his Great Uncle Wiley Trout, and ended his thoughts by saying, "My column does not go very far, nor amount to much after it gets there." Now, I can sure identify with that. But even more significantly, the title of this particular essay describes my column to a tee: "It Is Bad Some Days, Worse on Others."

***

One of the most cherished items a boy could possess when I was growing up was a brand-new Barlow pocket knife.

You can bet that any member of our group who obtained one was mighty eager to let the rest of us know it. If there were no slingshot forks to be cut or no cedar chunks to be whittled, the proud owner would nonchalantly clean the dirt from under his fingernails any time he could draw a crowd.

We didn't know it then, but the old Barlow, with its two folded blades, had been around for a long time and was even referred to back in 1876 by Mark Twain in his Tom Sawyer and Huckleberry Finn stories.

The primary use that most of us coal campers found for a Barlow was a good game of mumbly peg, which was a lot harder on the knife but much less strenuous to us than most of the other games we'd play, and usually occurred in the cool shade of a big elm.

The only skill required was the ability to stick the pointed blade into the ground via various "feats," probably 10 or 12 of them, all told.

The feats included things like taking the point of the blade between the first and second fingers, holding the hand toward the ground and flipping it with a jerk so the knife turned over once in the air and hit the ground blade first. The further into the game we got, the more difficult the feat, like holding the knife by the blade in your right hand, reaching across your chest and touching your left ear with its handle, then touching your right ear with your left hand and flipping the knife, causing it to rotate and stick in the ground.

As complicated as all this is when you see it written, it was really quite simple once the first player did it. All you had to do was what the person in front of you did, and usually after watching someone else do it, it wasn't all that hard to learn.

Anyway, when one player had successfully completed all the different feats, which usually ended when the winner stuck the knife into the ground by tossing it backward over his head, he'd take a peg (usually we'd use one of those big kitchen matches) about two inches long and, using the handle of the Barlow, drive it into the ground. The winner could take three whacks at it with his eyes open and three with them shut. Then the loser would have to pull the peg from the ground, using only his teeth. Sometimes, if the peg had been driven good and deep, the loser would have to eat a lot of dirt and everyone would have a good laugh at his expense.

I guess we could have used any kind of pocket knife to play mumbly peg, but I can't remember ever playing it except when somebody got a new Barlow.

***

Something we learned about rather early in life was rabbit backer. In the late fall we'd start our annual harvest. For those not familiar with the product, rabbit backer ("backer" being our pronounciation for tobacco), a.k.a. life everlastin,' was that little weed from which we'd strip the leaves to roll in a piece of brown paper poke to smoke. When no poke was available to roll one, and sometimes even when it was, we'd just chew it.

A few years ago a reader of my newspaper column sent me a clipping from an out-of-state newspaper about that same little weed, which I found to be extremely enlightening. The author of the article reported that he'd read in one of his wildflower books that rabbit backer was also known on other parts of this planet by a variety of names, including "catfoot," "poverty weed," "old-field balsam," and "cudweed."

I suppose we Muddy Branchers might have been receptive to calling it "poverty weed" since that's what we were living in, although I doubt we knew it. It's likely, too, that had we known, we couldn't have cared less. As I've said many times, we never missed what we'd never had and were no better or worse off than anybody else we knew.

And, who knows? If LBJ hadn't come in here in the 1960s and declared war on it, many of us might still be enjoying our poverty, as well as our rabbit backer.

Anyway, we might have even agreed to call rabbit backer "old-field balsam," which sounds like something you'd read about in the *Farmer's Almanac* and which we'd, no doubt, have pronounced "oil-field bomb." But I sincerely doubt we'd ever have resorted to calling it "cudweed." It would have reminded us too much of whatever it was the old cow chewed. I'm sure that, being the clever bunch we were, one of us would have concluded that it was named thusly because old Bossy slobbered or expectorated on it ... or even worse.

But thanks to that alert reader and the newspaper clipping, if we're ever talking to anybody from somewhere else and they mention that they once chewed "catfoot," at least we won't be conjuring up some sort of perverted image.

***

Stretching all the way across the front of the room, the old school blackboard was undoubtedly the most-used educational tool ever invented.

As soon as the teacher entered the room, snatched up the eraser and hurriedly removed all the hearts pierced with Cupid's arrows and filled with the Bobby plus Susie Maes; wiped away the crudely drawn stick figures under which somebody had written the word "Teacher," the blackboard was transformed into a virtual encyclopedia, providing the roomful of bright-eyed coal-camp kids with everything they needed to know.

Risking unimagined peril, the teacher would turn her back to the group and list the day's five spelling words in the upper left-hand corner, the arithmetic assignment square in the center of the vast space, and the nine names (written small enough to leave room in case others could be added if necessary) of the boys who had to stay in at recess.

All the time the teacher was writing, we sat in anticipation awaiting the instant her fingernail would accidentally rake across the smooth, slate surface and make that familiar screech that would cause all the girls to shriek in pretended agony and complain of the cold chills that ran up their spines. I guess it was sort of the same feeling as when you pulled a popcicle stick through your teeth.

The blackboard, or at least the soft, messy chalk the teacher used to write on it, was also used as a yardstick to measure just who was and who wasn't the teacher's pet. The two or three "pets" always got to leave the room a couple of times a week so they could go outside and "clean" the erasers. Actually, it was a job I wouldn't have particularly wanted anyway, since all they did was stand out in the schoolyard somewhere, hold two erasers at arm's length and slap them together until those slapping them could breathe again without coughing.

Quite naturally, since kids seldom thought beyond their most immediate wants and needs, we never gave much thought to what it might have been like at school before Mr. George Baron at the West Point Military Academy introduced blackboards and chalk back in 1801. We never considered the fact that the teacher would have had to write each spelling word 15 or 20 times, depending upon the number of pupils in the room, and the considerable amount of time all that would have taken; that it might have even cut into our recess and lunch periods. Mercy!

Research tells us that blackboards were made of slate until about the 1960s. Steel boards coated with porcelain enamel were used then, and to ease the starkness afforded by blackboards, green boards were introduced as an alternative. Even though they're green, most of the people I know still call them blackboards. Eventually, people started to refer to them as chalk boards, but now, teachers don't even use chalk anymore. In the

Coal-Camp Chronicles

late 1990s, someone came up with the dry erasable, white board where the teachers use colorful markers instead of chalk.

Losing the blackboard is all in the name of progress, of course, but one has to wonder how, in this day and time, the rest of the class can tell who the teacher's pets are.

***

Smelling musty from having been long stored in the darkest corner of my closet, my high school yearbook — being needed for some research for a newspaper article I was preparing — was

The brigde crossing the Levisa Fork of the Big Sandy River at Greentown collapsed on April 8, 1952, killing the driver of a truck hauling heavy equipment.

dug out. Sure enough, like an old friend you can depend on, I quickly found the information I needed. But perhaps even more importantly, I also discovered that those pages conjured up some long-forgotten memories; memories of days when worries were few and the future stretched before us like an unmarked highway, just waiting to be traveled.

It was August 1953. I was 14 years old and scared to death as I stepped for the first time onto the big yellow school bus driven by Curt Meade and bound for Tutor Key, River, Offutt and Williamsport, site of Meade Memorial High School. Ordinarily, the bus would have been pointed in the opposite direction when I boarded it, but the route that year ran backward because the bridge that spanned the Big Sandy at Greentown had collapsed on April 8, 1952, killing the driver of a truck hauling some sort of heavy equipment.

Anyway, the kids who caught the bus at the mouth of Boyd Branch had finished the eighth grade at the H. S. Howes Community School at Muddy Branch, and like me, were continuing their education at Meade. The bus trip itself would prove to be a kind of adventure as every day that we went to school we'd have to get off the bus at River and walk across the swinging bridge and get on another bus on the Offutt side. I remember vividly how on the last day of school, some of the older students, as they crossed the swinging bridge for the last time, donated their textbooks to the fish below. In those days, students had to buy their own books, preferably second-hand. So, in many cases, the books were so tattered and worn that they couldn't have been re-sold anyway.

The first time I walked inside the white, block structure that sat near the confluence of Greasy and Two-Mile Creeks, I marveled at its vastness ... its cavernous hallways lined with freshly-painted lockers and large photographs of former Red Devil basketball teams. It was the largest building I'd ever been in and even though I was terrified, I was filled with pride in knowing that I'd soon be an official member of its student body.

Meade Memorial High School as it looked in 1957.

I hadn't stepped fifteen feet inside the hallway until I was approached by a tall, balding gentleman wearing a tan suit and brown tie who asked, "And what is your name, young man?"

I managed somehow to tell him, whereupon he then told me who my parents were, as well as the names of all my brothers and sisters who had attended — and graduated from — the school in previous years. The gentleman was, of course, the legendary George W. Butcher, a highly respected educator who had been a member of the school's first faculty, some 22 years earlier.

George W. Butcher

I guess those old yearbooks (when did we stop calling them "annuals?") hold more than just pictures of pimply-faced kids with funny haircuts. It seems there's also a memory or two locked inside every page.

***

When I began my 33-year-long career as an educator back in 1961, I was hired to teach at Meade Memorial School. I had graduated there a mere four years earlier and many of my old high school teachers, like Helen Elam, Clarence Dutton and Ruth Salyers, were still plugging along, trying to stuff some learning into noggins like mine. It was very awkward at first, even surreal in some cases, and very difficult for me to dare think of myself as their peer.

To their credit, however, they accepted me totally, figuring, I guess, that having earned a bachelor's degree from Eastern Kentucky State College during the four years I'd been away qualified me to be counted among their number. Anyway, I was the first art teacher in the history of the institution and as such was charged with introducing the world of art to the children of the Williamsport, Boons Camp, White House and Offutt areas of Johnson County. Of course, high school students came from other communities throughout the county, but students from the communities just mentioned pretty much dominated the school's enrollment.

I had a couple of high school art classes every day and was the elementary art teacher. That meant that at least once a week, I had every elementary student in school in class. One little girl who was in the second or third grade, and whose name has long since left me, loved the color green. Regardless of the medium being used — crayons, tempera or construction paper — she always grabbed the green.

I recall once when we were painting landscapes, better known to the class as "outdoor pictures," she insisted that everything in her picture was to be painted green. The grass was green, the trees were green and the sky was green. No amount of instruction and suggestion from me could convince her to paint her sky blue.

Now, more than 40 years and hundreds of students later, I find that this creative little girl might have been right all along. Just maybe the sky *is* green.

What prompted both this memory and my sudden change of heart is based upon an article I read in the paper regarding an astronomy professor at Johns Hopkins University in Baltimore, who reported that the universe is indeed a pale shade of green.

Now, since my non-scientific little brain thinks of the sky and the universe as being basically the same thing, I can't help but feel just a twinge of guilt for trying to encourage that little girl to choose other colors for her painting.

I wish I could remember who she was then and who she grew up to be. I'd like to clip the aforementioned newspaper article and send it to her. Perhaps she can understand it even if I can't.

***

Back in January of 2004, Mike Preston, a former student of mine who now owns and operates Preston Funeral Home in Paintsville, was showing me a yellow, brittle copy of a 1955 edition of *The Paintsville Herald* that someone had found and passed along to him. He wanted me to see that particular paper because it contained an honor roll list from Meade Memorial High School, and lo and behold, I was on it.

A few days later, I received an e-mail from daughter-in-law Marcy (then a copy editor for the *Orlando Sentinel*) regarding an Associated Press story about honor rolls. It seems that Nashville, Tennessee, schools had stopped posting honor rolls (apparently on advice from school lawyers) because doing so might be a source of embarrassment for those kids whose grades were too low to make it. Some schools in the district cut out academic pep rallies and may even decide to cancel spelling bees and the like because the underachievers might get their feelings hurt.

Of course, it all has to do with the fear of being sued over privacy laws and such which some think forbids releasing academic information — good or bad — without permission. The Nashville schools were attempting to send out permission slips so parents could give the okay for the schools to tell the world that little Johnnie had a B average.

I've no idea how I managed to live with the embarrassment for all those other times an honor roll was published without my name being on it during my four years at Meade. Then again, it could be that all those times I wasn't on it just might have made me want to work a little harder so that the next time I would be. Guess that's just an old-fashioned idea that went out the window along with the idea that the teachers were in charge of the schools.

***

"You're all imps of Satan and doomed for perdition."

Those were the favorite words of one of my high school teachers whenever her students acted like ... well, imps of Satan.

She didn't utter those words every day, but found a need for them often enough for them to linger in the far recesses of my brain half century later. Even though I heard it often over the years I was in high school, I recall vividly the very first time I ever heard her come out with the soon-to-be familiar expression. It came just after lunch one day when the student body was all abuzz over one of the boys who had allegedly "taken dope" over at Walt Pack's store during lunch time.

What the young man had apparently done was ... gasp ... put two aspirins into his bottle of pop.

"He'll be as high as a Georgia pine," was the assessment of most of his classmates. "He's just an imp of Satan, doomed for perdition," was the aforementioned teacher's verdict.

As we now remember and laugh at what surely must have been an exciting distraction to our school day, we can only long for such days again. Days when the so-called "alkies" consumed shaving lotion and mouthwash; when street drugs and meth labs were so far in the future that mixing aspirin and pop was considered as bad as it could possibly get.

Of course, at least as far as this particular teacher was concerned, being an imp was not reserved for just dope heads. You were also an imp if you were inattentive in class, and especially if you were prone to "popping that blow gum," as she would put it.

Sometimes I got the impression that this teacher had let the expression evolve into a term of endearment; something she reserved for only the people she liked. For example, whenever Pikeville High School's basketball team ended Meade's up-to-then perfect season in the finals of the regional basketball tournament in 1955, no amount of coercion could prompt her to call the Pikeville players imps, which at the time was very difficult for me to understand since they so obviously were.

Anyway, her imps of 50 years ago have no doubt by now spawned dozens of children and grandchildren who have also earned that title. And wouldn't it be wonderful if we still lived in a society that would be appalled if a teenager popped an aspirin?

***

I found a couple of faded clippings from *The Herald* regarding the building where I attended grade school; the big yellow structure that sat majestically (at least it looked majestic to me) overlooking the school house bottom — that institution of lower learning that started half a dozen generations on the road to formal education — to be very enlightening. First of all, the clippings spoke volumes about the community itself; secondly, they provided haunting reminders of the time in which we lived.

The first, dated March 26, 1942, read as follows:

Friday has been designated as Victory Day at the Howes Community School at Thealka. Last Friday Verne P. Horne, superintendent of the Van Lear schools, gave a most valuable talk to the patrons, students and teachers at Thealka, on saving during the war.

The students of the school purchased $9.85 in saving stamps. Mrs. Edna Earl LeMaster's room, the fifth grade, led the school in stamp purchases, and her pupils will have a half day holiday Friday for leading the stamp sale.

A program will be worked out so that every teacher will be given a Victory Day chance at leading in the purchase of stamps. A drive is now on in the fourth, fifth, sixth, seventh and eighth grades for stamp sales. The room that wins will be given a picnic on April 3 for leadership, at the expense of the other grades."

Then on April 9, a follow-up article reported the following:

The Thealka elementary school is participating in the thrift plan of purchasing United States Defense Savings Stamps. The plan was adopted three weeks ago, and Friday of each week was designated as thrift day. On the three thrift days since the plan has been in effect, the students of the school have purchased a total of $42.60 in stamps. The students purchase the stamps from the Thealka post mistress.

The students are showing a fine spirit of patriotism, and plan to continue the thrift plan of purchasing throughout the year.

At the end of each month the school will submit to The Paintsville Herald *for publication a list of all students and the amount purchased.*

The faculty of the Thealka school are Mr. John Nichol, principal; Mrs. Edna E. LeMaster; Mrs. Nelle Corder, Mrs. Wm. Hazelrigg, Mrs. Lillian Arrowood, and Mrs. Garnett Ison.

In 1942 America was at war. Everybody — including grade school children in a tiny Eastern Kentucky coal camp — was pitching in.

***

Over the years, our little school often made the paper. Here are a couple of other examples:

The Paintsville Herald, September 29, 1927 *— The honor roll for the second month of school at Thealka School is as follows:*

Coal-Camp Chronicles

First Grade: Carmen Brooks, Curtis Castle, Avery Oakley Chandler, Ernest Castle, Irene Castle, Sue Belle McKenzie, Nellie McKenzie, Thomas Preston, Cecil Stapleton, Loman Stapleton and Alma Brooks.

Second Grade: Callie Preston, Mary Christine Burton, Junior Castle, Edna Mae Davis, Edward McKenzie, Jemima Jane Preston, Dorothy Robinson, Virginia Nelson, Orville Castle, Anna Mae Caudill, Glen Nelson, Charlie Sturgill, Virgil Stapleton, Wilma Davis and Garner Daniel.

Third and Fourth: Della Marie Greene, Esther Curiutte, Amanda Caudill, Golda Mae Castle, Carroll Davis, Alpha Brooks, Kathryn Castle, Bertha Brooks, Bertha Mae Preston, Bruce Castle, Carmel Castle, Johnnie Castle, Marcus Davis, Ernest James Davis, Cecil Preston and Vance Stapleton.

Fifth and Sixth: Pauline Mae Chandler, Mary Mae Welch, Thelma Greene, Rhoda Johnson, Helen Davis, Pauline Hunter, Gladys Collins, Josephine McClure, Fred Burton Whalen, Billy Boy Davis, Walter McKenzie, Ira Castle, William Robinson and Mildred Conley.

Seventh and Eighth: Oma VanHoose, Cecil Sherman, Harry Huff, Earl Davis, Ira Greene, John Burton, John Robinson, Viola G. Miller and Cecil Robinson.

***

The Paintsville Herald, October 25, 1928 — We are expecting a large crowd to attend the spelling match Thursday night. This is something new for this community and we feel sure you will enjoy yourself if you will only come. There will be no admission charge.

Miss Daniels and Miss Harris are preparing an entertainment for Halloween. There will be three dialogues, several drills and recitations. Come out and have a real Halloween time. And don't become frightened, even though witches, ghosts, and goblins in startling guise appear.

The time for our bazaar has been set for November 17. When you see these articles on sale be sure to buy something, and by doing so, help build a walk to our school building.

Pauline Hunter is back in school after a week's absence to take a trip to Alabama. She reports a very delightful trip, and has many interesting things to tell us about.

We are surely glad to have Guy Preston Jr. back in school. He has been absent a month on account of illness.

***

The Paintsville Herald, February 1, 1940 — The P.T.A. held its last meeting January 23, 1940. The following officers served the past year: President, Mrs. Dennis Daniel; vice president, Mrs. Katherine Dale; Secretary, Mrs. Edna Earle Lemaster; Treasurer, Mr. Eugene Daniel.

Five of my brothers and sisters are somewhere in this photo taken in front of the Muddy Branch school around 1939.

The P.T.A. of 1939 and 1940 has been the most successful Thealka has ever known. Some of the things that have been accomplished this year are, namely: installation of city water;

hot lunches for pupils; extension of the school term from seven to nine months.

The P.T.A. also sponsored a play, "The Last Day of School" Many improvements have been made in and around the school due to the efforts of this splendid organization.

The P.T.A. sponsored a play at the Thealka school. Cast members were, left to right standing, Billy Fraley, Emogene Burton, Ray Short, Ernestine Spears, James Lyons, and Bo Burton. Seated, left to right, Lizzie Colvin, Ernest Ray Preston, Frances Colvin Short, and Alta Lee Preston.

Coal-Camp Chronicles

Our principal, Mr. Ernest Rice, has built up the school and created such an interest that the people of Thealka feel they can look back on the past year with pride and forward to the year of 1940-41 with assurance of having the best school in Johnson County.

***

The Paintsville Herald, October 8, 1942 — The Thealka P.T.A. was organized September 25. Officers elected for the year were as follows: President, Mrs. Ernest Neal; Vice President, Mrs. Eddie Franklin; Secretary, Mrs. J. E. Corder; Treasurer, Mrs. Vencil Nelson.

The following were appointed by the president as chairmen of committees: Program, Mr. A. I. Lewis; Music, Mr. Ray Conley; Finance, Mr. Dan Gambill; Refreshments, Mrs. Leslie VanHoose; Ways and Means, Mrs. Prudence Castle; Membership, Mrs. Foster Burton.

The first meeting was held. Oct. 2 and an interesting program was given by the girls and boys.

1. Group singing; 2. Quartette: Wanda Miller, Ellen Jean Franklin, Ruth Dove and Ruth Aileen Castle; 3. Poem: "The Turtle," third grade; 4. Duet: Ellen Jean Franklin and Ruth

Aileen Castle; 5. Poem: "The Wind," Naomi Gambill; 6. Joke: Mr. Chandler.

Much enthusiasm is being shown at these meetings and it is hoped that the membership will represent every home in the district. The meetings are to be held the first Friday in each month at 2:15 p.m.

***

It didn't make the papers, of course, but something else which was special about the school house was the fall when two young men from town toted a 16mm movie projector and a cowboy movie across the long wooden bridge that connected the building with the rest of the world, and turned our auditorium into a movie theater. They showed up six or seven weeks in a row, tacked a bed sheet to the wall and treated us, at 15 cents a whack, to Johnny Mack Brown, Buster Crabbe and Colonel Tim McCoy.

In the days before TV, those were special nights for coal-camp kids.

Chapter 7
Rememberin' Stuff

June 1 is a special day for me.

It was on that date in 1961 that I marched across the stage of Brock Auditorium on the campus of Eastern Kentucky State College in Richmond and was handed a hard-earned BA degree by Dr. Robert Martin, the college president.

Also in line for a sweaty handshake was then-vice president of the United States Lyndon B. Johnson, who had just delivered a rather boring commencement address. Perhaps it was boring because I paid little attention to what he was saying due to the fact that I worried — actually I was 100 percent sure — that when it came time for my name to be called, it wouldn't be. It was, though, which made it O.K. that the diploma Dr. Martin handed me belonged to someone else. I retrieved my own from the dean's office minutes after the ceremony and went home happy.

But that event, as important as it was, is not why June 1 will always be a special day. Just two short year's later I was honored beyond belief when Wilma Jean Penix, daughter of Hobert and Almira Penix, of Williamsport, Kentucky, became my wife. To this day, I still don't know why a sane, intelligent woman would ever have agreed to do so, but she did. Truth be known, on many occasions during the interim, she's probably asked that same question at least a hundred times.

Over the years I've jokingly tried to convince her that I'm the best thing that ever happened to her — except, of course, son Todd and granddaughter Alison — while all the time knowing that it was I who had struck gold.

Guess that's why I got so aggravated the time several years ago when an emergency-room doctor at Cabell Huntington Hospital in Huntington, West Virginia, inquired as to whether the woman who stood beside the examining room table holding my hand was my significant other. In no uncertain terms I told him, "Absolutely not! She's my wife."

He didn't get it, probably thinking I had a hearing problem and didn't understand the question. But as she stood there smiling, I knew that she understood.

***

Although witnessed by hundreds of people, no photographer was standing by to snap that brief instant and record for posterity the time when the palm of my hand and the palm of a future president of the United States met in handshake.

Lyndon B. Johnson stood Texas tall as I held a bachelor's degree in my left hand and shook his hand with my right. Wish I could remember what he muttered as we briefly touched.

I gave little thought to that fleeting encounter until a few years later when he visited Paintsville as president and Wilma Jean and I stood on the sidewalk on College Street as he and Lady Bird rolled by waving at us from the back of a long, black open convertible. It didn't impress her much when I told her that I had once shaken his hand. Of course, it wasn't long until I forgot all about seeing him again until I'd hear something about his war on poverty.

Several years ago I saw the tail-end of a documentary on TV about "unforgettable characters" that had passed through people's lives, and some politician had commented on how LBJ had influenced him.

Guess I hadn't thought much about the word "character," much less about our 36th president being one. I suppose I always thought of a character as being someone with whom you were familiar; who was always a little different than most folks around; someone who, for one reason or another, set himself apart from everybody else.

To me, characters were folks from my growing-up days in Muddy Branch, like Granny Cotton who, when she milked her cow would go to the cow instead of having the cow come to her. I can still see her, milk bucket in hand and bibbed-apron flying, scurrying through the camp along about dark in pursuit of her wandering bovine.

Then there was Booten Puckett, who could make up a rhyme on a dime and remove an ugly seed wart from a little boy's hand by merely rubbing it, provided, of course, that little boy "believed."

And Liss Baldridge, my grandfather, was quite a character and could wear long sleeves and long underwear on the hottest day of summer and hardly break a sweat.

I suppose then, like the well-known *Reader's Digest* article suggests, everybody does indeed have "unforgettable characters" in their past. And since it's apparently legal to call a U. S. president a character, and since LBJ is a part of my past, he's in excellent company when placed alongside Granny Cotton, Booten Puckett and Liss Baldridge. Unforgettable characters, all.

Burns Mollett was another old gentleman who was as fine a person as you'd ever want to meet. Actually, since I'm much older now than he was then, perhaps using the word "old" might be inappropriate. Anyway, back in the early 1950s, all the kids in our particular neighborhood loved him even though some of the older folks often complained about his insobriety.

At least in my case, though, he is remembered for two things. One, he was a great storyteller, and two, he had a certain way with words that made simple, everyday conversation with him just downright entertaining.

One word he used often was "cattywampus." If something didn't quite suit him for some reason, he'd say it was cattywampus. As a pre-teen, I loved that little catch phrase and, perhaps thinking that my doing so might somehow make me appear as intelligent as I perceived this man to be, I used it as often as I could myself. I doubt that I ever used the term correctly, and I'd say things like "This arithmetic problem is cattywampus," or, "Ole Gene Autry sure is a cattywampus cowboy."

Recently I read about "cattywampus" in a national publication after all these years thinking it was just a home-made expression that Burns had invented. Matter of fact, it was kind of disappointing to learn that the term is used all over the country and probably originated more than two hundred years ago. Everybody everywhere likely knows it means "out of order."

Another word my long-ago friend used on a regular basis, probably so he wouldn't have the need to use regular cuss words, was "dadgum."

"This dadgum bicycle wheel is all cattywampus," he'd say.

It took some doing, but I also discovered that "dadgum" is a real word, too, is used everywhere, and also likely originated back in the 1800s.

Fortunately, my discovering that he hadn't invented either term has in no way tainted my memory of Burns, whose tales of his war-time adventures — told in the cool shade of his sister Lizzie Colvin's front porch on those faraway summer afternoons — kept a group of ragtag, coal-camp little boys spellbound for hours at a time.

However, if I someday learn that his heroic deeds while serving in South Africa didn't really lead to our winning WW II, now that might be another thing altogether. One thing's for sure, though. Those of us who hung onto every word he spoke on those occasions were convinced that ole Hitler was as cattywampus as they came.

***

Ever since I was a kid, I've heard older folks talking about how fast time flies as one gets older. Back then, however, that concept, which I now accept as pure fact, was terribly hard to grasp -- seemed like Christmases and birthdays were years apart.

But the good news as it regards the swiftness with which the world goes by these days, is that whenever the leaves fall, it won't be long before the snow does likewise. Then, we'll start hearing the frogs sing their little spring-time refrains and the grass will start turning green again. Fragile dandelions will literally take over our lawns and we'll bemoan the fact that the hedge is about to need trimming. It's always been that way, of course, but in the old days, it just didn't all happen so all at once.

In a 1940s coal camp, chilly weather not only forced us to once again start wearing our shoes, but, much to Mom's chagrin I'm sure, also drove us indoors. Fortunately, I had a little brother with whom to share those long, late-fall and winter afternoons. Older siblings kept us supplied with a ton of funny books and one of them had bought us a card game called Authors. (That's where I first heard of Louisa Mae Alcott, Nathaniel Hawthorne and Charles Dickens.) We had a Monopoly game, too, and although I don't think we ever completed a game, we spent hours on end buying hotels and passing "Go" and collecting $200.

Nevertheless, even with those distractions, in the days before TV, we still got bored from time to time, and, as would be expected, had a difficult time staying out of trouble.

For some reason, Mom would get all excited when a wrestling match would get out of hand and a perfectly executed flying mare resulted in our turning over the churn she had set next to our open fireplace to clabber; or when an aggressive re-creation of the Saturday-matinee cowboy movie's saloon-brawl scene ended up with a tablecloth filled with dirty dishes crashing to the floor.

The winter holidays were great, though, and Mom always seemed to make a special effort for Thanksgiving dinner. She had a super recipe for dressing that has been extremely difficult to duplicate. Of course, she did equally well on Christmas, and her cooking, topped with the presents we'd get — which more often than not included two new sets of Hopalong Cassidy cap busters which we were only allowed to shoot outdoors — managed to hold our attention while giving Mom a brief respite from chaos.

But as far as seasons went, my time was summertime.

Despite threats of bodily harm from both Mom and Dad if I were caught even near the river, sometime I still went there.

It was just after one of those times when the backwater had been several feet deep in the school house bottom that a bunch of us decided that since there was an ample amount of good-size planks left in the debris scattered along the creek banks, it might be a good time to build our cabin. Over the years, we probably started the cabin project at least half a dozen times. Naturally, since it would have required a bit of skill with saw, hammer and nail, plus a commitment of more than a couple of hours, it never got built.

Anyway, about half a dozen of us started pulling pieces of scrap lumber of all description from the banks of Muddy Branch and dragging them to higher ground. During this process, however, one of us noticed that the closer the river we got, the better the pickings. Near the Greentown culvert where the creek runs beneath the railroad tracks and Route 581, we even found a couple of planks about four feet long that appeared to never have been used for anything else. I suppose we just figured that if building material of this quality could be obtained from a measly little creek bank, it'd be untelling what might have washed up along the river.

"Only one way to find out," one of us said.

As we descended the bank just below the Thealka crossing, the forbidden Levisa Fork of the Big Sandy, as brown as a biscuit and still slightly out of its banks, sure enough appeared to have deposited an unending supply of good stuff. The thought briefly ran through my mind that our cabin would look like a mansion.

Only problem was, the high water had also left about a five-inch thick layer of mud along the shore and as both feet went out from under me, my bottom began an uncontrolled slide toward the fast-moving stream. I was completely helpless. There was absolutely nothing to grab on to. In the instant it took to process the dire straits in which I found myself, I knew I was a goner. I would drown for sure.

I must have been a comical figure as I slid quickly toward my doom because all my fellow lumber gatherers started laughing. Down, down I went. The slide must have been over in mere seconds, but it seemed forever before both my feet hit the water ... and a solid object that I later learned was a crosstie from the railroad tracks, which had lodged against some big rocks.

So there I sat. Although I then realized that I wasn't going to drown, I also knew that there was no way I could negotiate back

Coal-Camp Chronicles

up the slick river bank ... even if I could gain footing enough to turn around.

But, I suppose a body can do a lot of things if he has to ... and gets mad enough. The cheers and jeers of my cohorts, who apparently had failed to understand that I could have been killed if that crosstie hadn't stopped my slide, prompted me to somehow scramble on all fours back up the bank.

Or course, when Mom saw what a shape my clothes were in, there was no way I could deny where I'd been. As the Bible says, your sins will find you out.

***

It felt like Christmas back in the 1940s when one of us coal-camp kids would find an old tire. Just as soon as one would wear himself out pushing it wherever he'd dare to push, another would grab it and do likewise.

Unlike many of today's toys, old tires did not come with any kind of warranty or warning label. However, the latter might have been in order and would likely have read thusly: "If you put your little brother in this tire and roll him down the big hill in front of Jeff Sparks' house and he hits the wooden fence, he could get quite a few scratches and bruises about his face and head which might lead to a few red welts around your own boney little legs."

But in those days, an old tire or any other little thing out of the ordinary that we could use as a distraction, was eagerly acceptable as a toy.

As a matter of fact, many of us learned to sail a lid from an eight-pound lard bucket with greater accuracy than a lot of kids, and even some grown-ups, can with today's Frisbees. In defense of the Frisbee, though, I never ever saw a dog jump into the air and catch a lard can lid.

Lard can lids actually served another purpose as far as we were concerned because some of us learned to drive by running through the camp pretending that a lard can lid was the steering wheel of whatever vehicle we chose to be driving at the time. We even provided our own sound effects.

"Udden, udden ... screech!"

The "udden ... udden" part was when we were trying to make a steep grade and the "screech" was for when we'd go around a sharp curve or stop suddenly.

Then there was the bicycle rim. I never did quite master the art of pushing one of those things with a bent coat hanger like some of the older boys could. Guess I lacked the coordination and speed because after a few feet, the rim would simply run off and leave me.

I didn't dwell much on my lack of ability in that respect, but I did spend a lot of time wondering how little David of the Bible managed to drill ole Goliath right between the eyes with a slingshot that didn't even have forks. Of all the pictures I ever saw of the event, not one pictured David's slingshot with forks. But ours had forks (that we'd cut ourselves) and any one of us could have sent ole Goliath packing, especially if we'd been able to find a good smooth railroad gravel about the size of a quarter. Some of us were deadly shots and kept stray dogs and neighborhood cats in a state of deep distress.

Wonder what today's generation of kids would do if they awoke some morning without video games and such?

***

I read an article in *Parade* magazine once by Michael Crichton titled "Let's Stop Scaring Ourselves." In it he discussed fears he had as a young man -- fears that were magnified because they were publicized in various media to the point that they were looked upon as nearly certain to come about. Of course, none of them did.

Global cooling, then global warming; world famine; a world population explosion; and even cancer from power lines were among his concerns. Then there were the theories regarding danger from killer bees, saccharin, swine flu, and even Y2K.

He says he got over them, and although his fears were viewed from the perspective of a young adult, it reminded me of much earlier times and the things I looked upon with dread when I was a kid growing up in a coal camp. Like, for instance, the neighbor who enjoyed teasing all the younger boys by threatening to cut off our ears. I guess he was looked upon by others in the community as being just an ordinary, hard-working coal miner, but I just flat didn't trust him. Although I knew he was kidding us and didn't believe for an instant he'd harm any of us in any way, just to play it safe, I can't ever recall being within 20 feet of him.

Then there was World War II. I had no clue about how far Japan and Germany were from Muddy Branch. To me, a mere kid of four or five, Flat Gap or Blaine were as close, or as far away, as Tokyo or Berlin. Therefore, when Gabriel Heatter read the news on the radio about bombs being dropped on foreign cities, I feared a stray one just might fall on my house while I slept at night. I can't recall ever discussing that particular fear with anybody, but I do remember checking with Mom whenever a plane flew over to ascertain whether or not it was one of ours. She'd listen for a second and confirm that it was not the enemy and I'd return to my play and she'd continue to hang out her wash.

And finally, the thing of which I was most afraid, was lockjaw. One was destined to an agonizing death if he took lockjaw, and I was very cognizant of the two ways you could get it. The first was stepping on a rusty nail, which I did on a regular basis in the summer time. The antidote was to have Mom pour the wound full of turpentine, then wrap it with an old sock for a day or two, with me checking it regularly for red streaks.

The other way you got lockjaw was to go swimming with a cut, or even a chigger bite, during dog days. Since I had some sort of wound on my body most of the time, I didn't swim much during dog days.

Fortunately, as time marched on, not only did I survive all those dreadful fears, I think, at least to some extent, I even conquered them. Anyway, I still have both my ears, no stray bomb ever fell on my house, and I never got lockjaw.

***

"Political correctness," whatever that means, seems to be discussed a lot these days and just between you and me and the gatepost, it's about to border on silliness.

For example, one of the more recent things I read on the subject regarded a new high school in one of the southern states that was attempting to name a Blue Devil as the mascot of its athletic teams, a la Duke University, perhaps.

Anyway, a group of "concerned citizens," whatever that means, was holding a big protest rally claiming that to call a team the Blue Devils would be sacrilegious. Kind of like the big protest made a few years back with groups claiming the Atlanta Braves were racists because fans chanted and waved rubber tomahawks while cheering on their team.

The whole thing is rather amusing to me because I can assure you that back in the 1950s, religion and racism never once entered the minds of the students of Meade Memorial High School when the Red Devils played the Inez Indians. Red Devils and Indians, as far as we were concerned, were neither more, nor less, than two basketball teams that likely played harder against each other than against anybody else so they could own bragging rights to the few miles of Route 40 that connected Williamsport to the Martin County seat.

Sacrilegious indeed!

Williamsport has always been considered an upstanding religious community and the Old Friendship United Baptist Church is just across the road from Meade School. Yet, high school athletics and religion, at least to my knowledge, never once crossed swords. The two were always kept separate, at least philosophically, with the tales of David and Goliath often discussed right alongside the basketball prowess of the legendary Butcher family as folks sat around the stove at Walt Pack's store, located smack between the school and the church house.

By the same token, I'd bet a nickel against a hole in a doughnut that, even though there was actually an Indian tribe called the Inez, when the team won the 1954 boys' state tournament, not a person in the whole state ever once thought of Geronimo or Sitting Bull. An Inez Indian was a tall, skinny blond kid named Billy Ray.

It's hard to believe that any clear-thinking individual would think otherwise, let alone spend time and money protesting such an idea. Oh well, as Forrest Gump might have said, "Silly is as silly does."

***

When I overheard a lady at the grocery store tell another shopper that she had made two cobblers, my memory kicked in and I automatically thought blackberry.

As I stood in line to pay for what few items I planned to purchase, I momentarily revisited my youth and since blackberries were plentiful in those days and canning was something practically every woman in Society Row did, I decided there's just no telling how many quarts of blackberries my mother canned every summer. And of course, the good news was that most of them ended up victims of her cobblers.

I recalled that Dad (who, by the way, had picked about eighty percent of the berries himself, either high on the hill in the head of Well Hollow or maybe even over on Teays Branch) enjoyed his cobbler hot, right out of the oven with fresh milk poured over it. Personally, I preferred mine to sit and cool a while before I ate it. I'm sure Mom made cobblers from fresh-picked berries as well as from canned ones, but I can only remember eating cobblers in the fall and winter. Maybe that's because I was likely indoors while they were being baked and was able to enjoy the full aromatic pleasures that only a baking cobbler can emit. One thing's for sure, unlike the smell of brewing coffee, cobblers did indeed taste as good as they smelled.

Anyhow, the lady's comment in the grocery store is the reason you're now reading about blackberries and desserts.

Of course, Mom also made jams and jellies from blackberries, but cobblers, jams and jellies are not the blackberry's only useful purpose. Old timers, and as far as I know, even new timers, used (or maybe even still use) blackberry vinegar and blackberry wine for various medical ailments. One lady told me once that sipping blackberry wine on a regular basis would tighten loose teeth. Another elderly gentleman said he sipped blackberry vinegar to help his rheumatism.

But, like a lot of other Appalachian folk ways, blackberry picking, and even canning, are quickly becoming lost arts. I know for a fact that I was once a big-time blackberry picker and it's been nearly half a century since I've been blackberry picking, and probably that long since I've seen anybody else come strolling out of the woods carrying a big water bucket heaped full of blackberries.

***

I feel terrible about the role I might have played in the impending extinction of one of nature's creatures. But had I known he was merely 50 years or so away from the fate of the dodo bird, perhaps I would have taken a different approach to fun and games when I was growing up.

I'm talking about an article that appeared in the newspaper procclaiming that the waterdog — some folks call him a mudpuppy — with which I played on a near daily basis in the summer time is now "rapidly becoming threatened worldwide."

Who'd a thunk it?

Boyd Branch was full of them and if I turned over three rocks, a good-sized waterdog would be under at least one of them.

It's hard to imagine anybody who would read this book who wouldn't know what a waterdog is, but just in case, it's a type of salamander. The waterdog is about three or four inches long and relatively easy to catch. When I was a kid I learned that, just like an ordinary lizard, if you broke off its tail, it'd grow another one. Or at least that's what we thought and, come to think of it, I can't remember ever seeing one that didn't have a whole tail.

Some of the really avid fishermen who lived in Muddy Branch at the time would catch them during the day and use them for bait at night.

But now here's this newspaper article explaining that salamanders, and even toads, are becoming endangered. But let's make one thing perfectly clear: While I'll admit to bothering, and maybe even innocently abusing, the waterdogs that lived in my neighborhood back in the 1940s, I had absolutely nothing whatsoever to do with the fall and decline of toads. As a matter of fact, some of us boys may be the very reason that the toad has survived as long as he has. We would even risk a bad case of warts as we'd fill a toad with lightning bugs so we could watch them blink inside him. The lucky toad frog we'd pick for such a display probably wouldn't have to hunt anything else to eat for at least a week. So my conscience is clear regarding toads.

And bullfrogs? Same thing. I never went gigging in my life.

Anyway, here's this scientist from the World Conservation Union saying that nearly 33 percent of all the species of amphibians are about to become extinct. And I guess they threw in the fact that the same holds true for 12 percent of all bird species because they thought I'd feel guilty about all the times I'd hunted birds with my slingshot. But I've got them on that one. Truth be known, I couldn't hit the broad side of a barn with a slingshot. All that bragging I might have done about my slingshot prowess ... just brag, no fact.

I do feel bad, though, about those waterdogs.

***

Clyde Roy Pack

Thanks to way too many trips to the Sipp and Royal theaters and far too many Technicolor Indian uprising movies on Sunday afternoons, I managed to get some expert advice on just how to build a bow. After all, if the bow and arrow served Geronimo's and Crazy Horse's mighty warriors as well as they had, they were certainly good enough for whatever tribe I chose to join at any given time.

Although my personal knowledge of Native Americans came via Hollywood, as an occasional find of an arrowhead in Dad's freshly plowed garden or under a sheltered rock cliff would prove, Indians had indeed roamed the Johnson County countryside in the far distant past.

Anyway, I would pay my nine cents, sit in the dark and eat popcorn one kernel at a time while Iron Eyes Cody or Jay Silverheels taught me, and the half dozen other Muddy Branchers with whom I'd usually travel, that the key to making a good bow was to hunt a strong sapling to use. I doubt that I knew one tree from another back then, but there was a particular bush on the hill behind Edgar Preston's house that I utilized on a regular basis. I had sat through enough war councils to know that the longer the bow, the less likely it would be to break; something about there being more wood to share the load, whatever that meant. Anyway, I developed bow making into quite an art.

After I'd cut a little nick in each end of the stick that I'd selected for a bow, I'd bend it as much as I dared and string it with heavy twine that Mom would unravel from feed sacks. If I was really lucky, I'd come up with some black fishing line; it was much stronger. Again, taking a lesson from the movie Indians, I learned to keep my face turned away when I was stringing the bow. Still, sometimes I'd get slapped about the face if the string pulled loose or the bow broke.

For arrows, I'd gather up dried stick weeds. Although they would often stray from the intended direction whenever there was a breeze, they were perfectly straight; as a matter of fact, as I'd gather them, sometimes I'd hold up one for closer inspection and surmise that it was indeed, as straight as an ... arrow.

Nevertheless, despite their tendency to stray, I was pretty accurate with them at close range. I had to be careful, though, because I'd sharpen the ends so they would actually stick in soft targets and it would have been pretty easy to put someone's eye out with one of them. To my knowledge, no one ever got hurt by one of my bows and arrows -- at least not very bad.

Chapter 8

A Distinct Culture

I guess I was nearly grown before I realized that we Eastern Kentuckians had our own dialect. Like, for instance, my Dad would say that something was "lopper jawed" whenever it was lopsided. I've also heard him speak of the time when he was a boy and a "painter cat" was reported to be living in the hills of his native Martin County. A "painter cat" was a panther.

To his generation, if someone was very sick, he was "bad off." If he recovered, then got sick again, he had a "backset."

We still hear those kinds of phrases, especially from older folk.

It is not at all unusual for Eastern Kentuckians to pronounce single vowels as diphthongs (double vowel sound): *bad* becomes *ba-ud*, *end* is pronounced *i-und*, and *door* is often pronounced *do-ur*.

We're also known for pronouncing words ending in *ow* as if they ended in *er*. Folks in these parts live up a "holler," not a "hollow." And, he is a fine "feller," not a fine "fellow." We're also notorious for dropping the *g* at the end of a word and end up *goin'* and *comin.'* We also still have a tendency to change the *a* to *y* when we have us a *sody* pop. However, we've just about lost the habit of putting an *a* before a verb. It's been a right smart bit since I've heard someone say, "He's a-goin' home." It's also been a while since I've heard anybody say, "It's been a right smart bit."

Then you hear things like, "Bobby Joe is turned like his daddy, but he favors his mother's side."

Believe it or not, there are people — highly educated people, at that — who could hear that statement and not have a clue as to what it means. And, as amazing as it sounds, those same people would likely also draw a blank if you asked them to reach you the salt shaker.

Of course, we coal-camp kids understand perfectly that the first sentence simply means that Bobby Joe has a personality like his father (i.e. quiet, unassuming), but his physical features (that red hair and buck teeth) resemble those on his mother's side of the family. And for some reason, non-Appalachians are slow to grasp that "reach me the salt shaker" is our way of requesting that someone please "pass" the salt.

It's little wonder, then, that so many folks from "off" have such difficulty, not only with our everyday conversation, but also with some of our old sayings — superstitions, if you will — that we know as absolute truth.

For example, when it comes to predicting the weather, we know with 100 percent certainty that if it rains on the first day of July, it will rain 17 days in the month; if your corns itch it's going to rain; the number of fogs in the summer indicates the number of snows in winter; and a rain on Monday means two more days of rain that week.

We know that. We've *always* known that.

Just like our grandfathers and their grandfathers before them knew that if they planted beans in the morning, they came up two weeks sooner than if they'd planted them in the afternoon; that St. Patrick's Day is a good time to plant potatoes; and that turnips should be planted on the tenth day of August, except, of course, unless the tenth day of August should fall on a Sunday.

It's absolutely beyond me why anyone would have trouble believing some of the stuff that the rest of us have known since we knew anything. Like it's bad luck to kill a cricket; eating chicken gizzards will make you beautiful; and if you find a horseshoe, spit through it, throw it over your left shoulder and you'll have good luck. Surely they realize that when two people speak the same word at the same time, they should join their little fingers and make a wish.

And finally, one has to wonder how those born elsewhere ever got a sweetheart, much less a husband or wife, without knowing that if you take the last piece of pie on a plate, you will get a handsome husband or a beautiful wife; that if you can eat a crab apple without frowning, you can get the person you desire; or that anyone who can make the first and fourth fingers touch over the back of the others may marry anyone he or she chooses.

And how about the following recipe for wart removal?

Walk into the road at night when the moon is shining. Run around in a circle three times, then spit over your right shoulder.

Or, if you're afraid one of your neighbors might see you and call 911 to report there's a crazy person loose in the neighborhood, just pull a big tick off your dog.

Either way — if you believe in folklore — you're bound to rid yourself of warts. Of course, after you get the tick, you must let it bite the wart, then put it back onto the dog. A mere two or three days later, the wart will be gone.

Warts must have always been a plague on society because it seems that I always hear of more cures for them than for anything else. Another suggests cutting a section of bark from a tree (no particular tree was mentioned, no particular size of the section) and rubbing it on the wart. Then the section of bark was to be taped back to the tree. By the time it grows onto the tree again, the wart will be gone.

Don't care for either of these cures?

Try something a little more simple, like tying a horse hair around the wart, or putting butter on the wart and letting a cat lick it off. Or even better, bathe the wart with water in which potatoes have been boiled.

***

In the summertime, when all the leaves are green, the redbirds sing, and the skies are blue ... people got poison ivy, especially if they lived in an Eastern Kentucky coal camp.

When I was a kid, it seemed as if poison ivy would come to the house, knock on the door and look me up. By summer's end, I'd no doubt have had several bouts with the pesky stuff.

However, even though I had some pretty nasty cases, I can't ever remember having to go to the doctor with it. While I don't remember any particular cure that Mom used, I suspect she had stored away somewhere a few of the old home remedies like those just mentioned. Fortunately, we don't have to rely on them in these modern times, like rubbing garlic or the juice from a green tomato on the affected area to relieve the itch.

I've heard that people once used to cure the unwanted rash with lather from home-made lye soap; a mixture of molasses and baking soda; a mixture of Epsom salts and water (or buttermilk); juice from a milkweed; ordinary white shoe polish; slobber from a suckling calf; or rubbing the itchy spot with a fresh banana.

Looking back, one has to wonder what healing properties were contained in stuff like garlic or lye soap or milkweed. At the time, though, what difference did it make what made it work, as long as it did?

***

It is probably because I've written a book (*Muddy Branch: Memories of an Eastern Kentucky Coal Camp*) about growing up in a coal mining environment, but about once a week, or so, somebody asks me a question relating in some form or fashion to coal mining. Most of the time they seem surprised when I don't know the answer.

"I thought you were raised in a coal camp. How come you don't know that?" they say.

Well, while it's indeed true that there has never been an individual who was more of a coal-camp kid that I was, the truth is, when I lived in a coal camp, I *was* a kid. Much to the delight of my father (who loved mining but didn't want his sons to be miners), I've never set foot inside a coal mine, except once, and then no more than 25 or 30 feet from the entrance. As I've said many times, I knew nothing about mining and the industry itself, except what I overheard from conversations Dad would have with Mom or some of the neighbors. I was much too busy doing my own thing — trying to get enough boys together for a baseball game or a swim in the Number One Pond — than to concern myself with things that were best left to the grown-ups.

Oh, I knew what a carbide lamp was and a slate dump and a scab (both varieties) and I knew that John L. Lewis was president of the United Mine Workers of America. But although I heard terms like *bug dust, cap rock,* and *drift mouth,* I had no clue as to what they referred. To me they were just words that miners used when talking to other miners.

So, at the risk of being laughed at by the old-time deep miners with whom these terms have long since become as familiar as the backs of their hands, for the benefit of those who've asked, as well as those like me who didn't know but should have, here's a little glossary of terms we've heard all our lives without any idea as to their meaning.

Coal-Camp Chronicles

• *Bug dust* -- fine particles of coal dust. I knew one old coal miner who was given the nickname "Bug Dust."

• *Cap rock* -- a layer of rock mixed with coal that is located between the coal seam and the solid rock roof of the mine.

• *Drift mouth* -- walk-out opening from a mine to the outside.

• *Fireboss* -- foreman who inspects the mine for gases.

• *Mantrip* -- rail cars that carry miners from the outside to the work areas. "Mantrip" was the working title of Cheryl Ladd's movie (I was in it for about two seconds) filmed in and around Paintsville in the summer of 1981. They ended up calling it "Kentucky Woman" and it had to be the worst TV movie ever made. I enjoyed watching it, though, because I wanted to look for all the people I knew.

• *Pillar* -- coal seam left as roof support after mining.

• *Rock dust* -- powdered limestone, sprayed throughout the mine to eliminate the explosive qualities of coal dust. Actually, it's the rock dust that causes miners to develop silicosis, commonly called "black lung."

• *Roof bolter* -- a miner (or a machine) that places bolts in holes to support the roof.

I guess, then, as a kid, I made up for not knowing much about the coal industry by being pretty knowledgable about how to entertain myself. I was reminded of that on a recent winter afternoon when I stood at our picture window for a few minutes and watched some teenagers playing on a little slope with a big, brightly-colored, store-bought plastic object that loosely resembled a sled. Although the temperature hovered in the mid-20s, they seemed to be thoroughly enjoying themselves and it all reminded me of the snowy Saturday mornings 50 years ago when the coal-camp kids in Silk Stocking Row would spend hours engaging in a similar activity, except our choice of conveyance was made of pasteboard, not plastic. (For those under 50, pasteboard is what cardboard used to be called).

I guess those days are gone forever. In the first place, when was the last time you tried to scrounge up a pasteboard box? They simply can't be had.

I don't mean the kind you could stick under your arm to tote a dozen Manila envelopes and a few sheets of typing paper from home to work and back. I'm talking about the kind that three or four of us could get into at the same time and slide all the way from the top of the hill to the road; the kind that would eventually get soggy, come apart and, as ashamed as I am to tell it now, become a part of the landscape until springtime greenery would come and hide it from view.

Coal-Camp Chronicles

By and large, winter and I did not agree on much. As a matter of fact, as far as I was concerned, allowing me to slide down a snowy hill on a pasteboard box was about its only redeeming feature.

Of course, the wildest ride I ever experienced was not on pasteboard, but came via an old refrigerator door that someone found and decided would be fun to ride. It wasn't. Had I not bailed out only seconds into the ride, you'd probably be reading someone else's ramblings right now. The thing ended up in a neighbor's front yard ... after it'd taken out the bottom plank of his fence.

***

Something else at which I was pretty knowledgable was coal-camp cuisine.

"Pinto Beans" is what was printed on the package, but "soup beans" is what really came out of the pot.

To this very day, I'm convinced that the best way to eat them is to put them in a large teacup or soup bowl and crumble cornbread over them. They absolutely must be washed down with milk — the colder the better, and buttermilk whenever possible.

And let's make it perfectly clear, it's "soup beans," not "bean soup," like some of the fancier restaurants prefer to call them.

If there were ever a particular cuisine for which I could ever be considered an expert, this would have to be it. Of course, the same thing could probably be said of practically every other Appalachian who grew up in the forties and fifties.

Simply put, soup beans were a staple in those days. If a family was out of soup beans, it was pretty much out of groceries.

What brought on this line of thinking in the first place was the following question: "Did you ever mash mustard in your soup beans?"

Believe it or not, someone once did pose that little interrogative to me. My answer, of course, was, "Certainly. Who hasn't?"

When one ate soup beans practically every day of the week, he couldn't help but seek out a little variety. Since there was always a jar of mustard handy, the combination was a natural.

But soup beans were seldom eaten by themselves. For example, Dad enjoyed a hunk of onion with his. Note the word "hunk," not "slice." He'd peel and quarter a big onion, more often than not one he'd grown himself, and eat it much like one would an apple. He also liked a salt pickle (one that Mom had canned) with his soup beans.

Although I've heard others talk about their moms making soup bean dumplings, mine never did. But another bean combination that does come to mind from those suppers of long ago, is the mixture of pinto beans and white (or Navy) beans. I've no clue as to why we called white beans "Navy beans," but I do remember song lyrics from the radio that said, "The Army gets the gravy and the Navy gets the beans." Perhaps that had something to do with it.

***

Sometimes I can't even remember what I had for breakfast, but I still have vivid memories of walking the halls of dear ole Meade Memorial High School, wearing a butch-waxed haircut, skin-tight pegged jeans, a button-up shirt with the collar turned up behind (like Elvis) — and how could one ever forget — a face full of red and yellow bumps.

Kids today call them zits. To us, they were simply pimples.

Regardless of what we called them, though, thanks to earrings on boys, and body piercing and tattoos on both sexes, pimples are about the only things that our generation and the new generation have in common. And, I'll bet you a nickel that kids today are just as aggravated by them as we were.

I was always made to feel that having pimples was akin to having lice or something; that it was my own fault that my skin was oily; that they were caused by something I did (or didn't) do. I resented boys (usually dark skinned, black hair) who didn't have acne but envied the girls who did because they could cover them with pancake makeup.

All through my high school years, I had acne so bad that in the mornings when I'd shave (which I did at least once a month) I'd take the hide off the top of so many pimples that I'd look like I'd been shot slap in the face with a double-barrel acne gun. In order to stop the bleeding, I'd dab dozens of bits of tissue paper on the wounds and leave them until I was about ready to go catch the bus. Before Curt Meade would pick up the crew from Muddy Branch and haul us across Two-Mile to be educated, most of the time, I'd remove all the bits of red-dotted tissue. But if I overlooked one, some smart aleck would yell, "Hey everybody, looks like Clyde put a blade in his razor this morning."

***

I was five months shy of three years old when the Japanese bombed Pearl Harbor: too young to know about such things and what that fateful Sunday morning would ultimately come to mean.

Some of my earliest memories regarding World War II came a few years after that and include those of Dad and Mom listening to the evening news on the radio as Gabriel Heatter would sometimes list the names of American ships sunk by the Japanese in the waters of the Pacific. They, of course, were listening attentively, and hoping — make that praying — that the names of the ships that carried my two oldest brothers, Ulysses and Hubert — both of whom had joined the Navy as soon as they graduated from Meade Memorial High School — would never be heard. Fortunately, they were not. Both brothers returned home safely after the war.

Then there were the letters that came in envelopes with little red, white and blue stripes around the edges that cost six cents to mail; letters that were received after having been censored (part of which had been blacked out or completely cut away) by U. S. officials who feared military secrets would be revealed unknowingly by servicemen writing back home -- kind of like the "loose lips sink ships" thing. I can remember wondering, even at that young age, who it was that had read that letter and decided to black out so much information.

Neither of my brothers talked about their war experiences, at least not in my presence. But as I grew up in the coal camp during the years that followed World War II, just like most of my peers seemed to be, I was fascinated by war stories. Mostly they came via John Wayne in films like "Sands of Iwo Jima," or "The Flying Leathernecks," but seldom from those who actually participated in the conflict. I watched and listened in wonder to tales of places that sounded strange and were difficult to spell; places far away; places that Old Mrs. Cotton (one of our neighbors who lived to be well over 100) referred to as being "across the water."

By the time I was about 10 years old, the war was long since over. To us coal-camp kids, it was a memory of something glamorous; something to be mimicked with hours of war games among the rocky hills of Thealka. War was exciting ... and it was foreign.

Of course, some 52 years later, on September 11, 2001, the war came home.

It's one of those occasions where you'll never forget where you were when you heard the news; an occasion like when

President Kennedy was shot; when the space shuttle exploded; except worse in so many ways.

In the weeks, months, and now years, that followed, our faith has been tested like never before. We have been asked to make sacrifices unparalleled in our nation's history. We have found that the price of our freedom has been made abundantly clear.

Of all the things I've heard or read since the airliners were crashed into the World Trade Center and the Pentagon, none, in my opinion, better describes our situation than the quote by Thomas Paine: "Heaven knows how to put a proper price upon its goods; and it would be strange indeed if so celestial an article as freedom should not be highly rated."

As with the cowardly acts of Pearl Harbor, what happened in New York will be addressed. The difference now is that we won't be counting on John Wayne to tell us the story. We are living it day by day, week by week, year by year, and probably will as long as we live.

Chapter 9

Church and Such

I have no old attendance records on file to prove it, but I'll bet that back in 1950, I attended Sunday morning worship services at the Thealka Free Will Baptist Church 75 or 80 percent of the time that the doors were opened. About the only thing that could have kept me from going to church was some sort of communicable disease or extremely bad weather.

Sometime during my growing-up years, I did indeed have the measles, mumps and chicken pox, but at the same time, I can also remember going to church when we'd have to walk the main road out of Silk Stocking Row to the railroad tracks, then walk the tracks up to the church house because the snow was so deep (or it was so slick) that we couldn't take the short cut around behind Foster Burton's, Don Fraley's and Edgar Preston's houses. I can never remember a time when church services were cancelled because of bad weather.

What got me to thinking about all this was an article I read in the paper about some poll being taken that indicated church attendance is down to an all-time low in some parts of the country. All the obvious and expected reasons were listed, but few of them would likely have applied to the society in which Appalachia existed in those days.

First off, I'll admit that my church-going had little to do with my feeling of righteousness, or anything like that. While I'm sure it didn't hurt me any to be part of the congregation and be exposed to sermons that could make you feel the fire and smell the brimstone, my being there had more to do with entertainment than salvation. Sort of like, go to the movies on Saturday, church on Sunday morning.

I especially enjoyed the songs of Clyde Conley and the Tom's Creek Quartet, as well as other singing groups who'd visit us from time to time. Another memory of those days regards a little-boy preacher. I refer to him as that because he was even younger than me. I can't remember his name (I only saw him once) or from whence he came. Neither can I remember any specific thing he said, just that they passed a hat and took up an offering to help pay his travel expenses. But he did sound exactly like the older, old-time Free Will preachers that I'd heard all my life. As a matter of fact, I remember thinking at the time that this boy did an excellent imitation of a real preacher. I've often wondered if that was really what he was.

But even though Dad was a deacon and would have insisted that I fill a pew on a regular basis anyway, going to church when I was a kid was something I really enjoyed doing. Even after sitting up until one o'clock on Sunday morning to watch wrestling from Dayton on Bill Hampton's TV, more often than not, within eight hours or so, it was Sunday school and church for me.

***

I don't know how far its sound would reach, but you could sure hear that church bell loud and clear at my house. When the tipple was idle and no sound of any consequence was present on those long-ago Sunday mornings; when the familiar clear, rich peal from the bell tower of the Thealka Free Will Baptist Church echoed through the little valley and bounced off the Boyd Branch hillsides, it was like even the dogs stopped barking and the birds stopped singing long enough to listen.

When I got up to some size, I became part of the bell-ringing ritual myself as I proudly accompanied Dad, whose job it was to ring that bell thirty minutes before services were to begin, then again just as they were ready to start. He'd never let me pull that long seagrass rope that was looped around a large nail just inside the entrance, but I'd stand next to him and stare at the ceiling to see the movement that sent out the four-or-five-minute message to the whole community. I'd then — except in the wintertime, of course — park myself outside on the top step and watch to see if anyone had gotten the message. Without fail, within a minute or two folks would start drifting, one or two at a time, toward where I was sitting. The women, all smelling of fresh powder, Juicy Fruit gum and Evening in Paris cologne, and often with a kid in tow who would obviously rather have been anywhere else in the world, would quickly enter the building and take their seats. The men, on the other hand, would loiter outside for a few minutes, smoke one more cigarette or give one more chew to the wads in their jaws before spitting them out into their hands and tossing them away.

Somewhere along the way, I heard someone say that the big bell that alerted the worshipers to the little white clapboard church house and still rings in the far recesses of my brain, had at one time been part of a steamboat that navigated the Levisa Fork of the Big Sandy River in the late 1800s. That has never been proved to my satisfaction, however, but it appears to be very likely since riverboats were certainly bountiful on the Big Sandy at that time, and one theory is the community of Thealka itself was named for one of them.

Anyway, at the time I was "helping" Dad ring the church bell, I was of the opinion that it was indeed the ringing of the bell that caused the people to come to church; that without the bell, preachers like Don Fraley or Charlie Bailey or James Lyons or Raymond Dale would have had to preach their fire-and-brimstone sermons to empty pews instead of a row of squirmy boys convicted by those fiery words, especially when the preachers would make eye contact, which seemed to be most of the time.

It's not likely to ever happen, but it'd be sort of interesting, to me at least, if some sort of poll was taken comparing church attendance with churches that still ring their bells and churches that don't.

It could just be that the theory of a 1940-something coal-camp kid just might have had merit.

***

Although I've written often about the baptisms that I remember from my boyhood, I've never discussed footwashing.

That's about to change.

First off, though, I'll admit that even though the Free Wills with whom I was raised were indeed footwashers, for some reason, Dad forbade me from ever attending one of the yearly footwashing services held by the local congregation.

Sadly then, the following story is based on hearsay. I will state, however, that those I heard say it believed it wholeheartedly and told it for the truth.

It's pretty much a given that back in the late 1940s, few of the rural Baptist churches had indoor plumbing, which quite naturally forced those in charge of the footwashing services to tote water from the nearest well or from a nearby creek. So, on this particular hot, summer Sunday in this particular church, several Number-Two washtubs were filled in anticipation of the large number of believers who would come from far and near to participate in the ritual.

While the service was proceeding nicely, two boys (probably about 10 or 12 years of age) decided they were bored with the whole thing, slipped quietly out a side door and parked themselves in restful repose on the shady side of the building, right under an open window.

A few minutes later one of the deacons inside decided it was time to empty a few of the pans of water, which he did ... by simply holding them out the window under which the two young men were sitting and turning them upside down.

Clyde Roy Pack

A traditional coal-camp baptism. The preachers are Charlie Bailey and James Lyons.

Fortunately, the noise inside the church was louder than the startled, obscenity-filled comments that came from the two newly baptized ... well, sprinkled anyway, young men left sitting in a puddle of footwasher water.

Then there's the footwashing story about another two young men, probably several years older than the two mentioned above, who in about 1940 were constantly hounded by their hard-shell father to give up all that drinkin' and swarpin' and start going to church with him.

On the footwashing Sunday in question, the pair arose earlier than their father, who they knew would be going to footwashing in an hour or so. So, with that in mind, they sneaked into his room, removed his finest pair of black socks that he'd tucked inside his finest pair of go-to-meetin' shoes, and filled them both about half full of soot from the fireplace. Then, just as quietly as they had stolen them, they stuffed them back inside his shoes.

Folks in that particular community still talk about how the man had gotten so mad that one of the boys ended up joining the Army and the other left about a week later for Wabash, where he got a job with General Tire. As far as anybody knows, neither of them ever went to church with their father. Don't know about their drinkin' and swarpin.'

***

Being reared a Free Will, I did indeed sit under some mighty hard preaching. Especially those times when Mom would sit between Joe and me; when distractions had been cut to a minimum; when there was little else to do but pay attention to the preachers, I often got the message.

I may have been a mere kid, but I knew exactly what preachers meant when they'd say, "You reap what you sow."

I knew they meant that if you planted corn, you'd get corn. If you planted beans, you'd get beans. I was never too sure what all that had to do with church, though, other than the fact that practically everyone there had a garden of some kind out back of their house and the preachers didn't want anybody to get mixed up and try to pick white half runners off a corn stalk.

I was reminded of the "reap what you sow" message when I read in the paper a while back about some scientists out in Los Angeles planting a little lotus seed from China and how surprised they were when in about four days, a little green lotus shoot appeared.

That part should have been no surprise to anybody, but the fact that the seed was 1,288 years old and was still able to produce, apparently surprised them a bit.

I was a bit puzzled about something else: the article didn't say the seed was over 1,200 years old, or nearly 1,300 years old. It said the seed was exactly 1,288 years old. That amazes me more than the fact that it still produced. Just how did they know exactly how old it was? I mean, did it come in a Ming vase labeled "Lotus seed, 716 AD?"

Anyway, one of the professors involved in the planting of the seed said, "This sleeping beauty, which was already there when Marco Polo came to China in the 13th century, must have a powerful genetic system."

Well, of course it does. Anybody who has ever been to church a day in their life knows that.

And it's not because some scientists at UCLA, or some other institute of higher learning, figured out some magic formula, either.

And that's exactly what those old Free Will preachers were trying to tell us more than 50 years ago. It doesn't matter if it's a potato or a peanut or a paw paw or a pear, you reap what you sow.

***

Throughout the history of the camp, the church has played an important role, sometimes even making the newspapers.

The Paintsville Herald, Dec. 15, 1916 — *Rev. L. F. Caudill, assisted by Rev. T. J. Collins, has just closed a series of meetings here in which much and lasting good was* *accomplished. Bro. Caudill is an able and conscientious minister of God. He is modest, congenial and always in a happy mood. Looking into his face one can see the love of God shining therefrom. His standard of religion is very high. He preaches a God of grace and mercy and flays mercilessly the idea that God ever forsakes his children. Although they may have grown cold and have backslidden, He is still calling and pleading with them to come back into the fold. He prayed God to deliver from prejudice and spiritual blindness those to whom the people look for guidance and illumination.*

The meeting was void of all noise of emotionalism, but the silent power and working of the Spirit was wonderfully present at the first service and increased as the days went by. We have not seen such a deep spirit of real victory and holy joy on God's children for years.

The singing was conducted by Bro. Glen Preston who also assisted Bro. Caudill very materially in the meeting.

Miss Oneida Howes and O. J. Williams were united in marriage at the home of the bride last Saturday evening at 7 o'clock, Rev. O. J. Carder of Paintsville officiating. Miss Howes is a beautiful and accomplished young lady and is the only daughter of Mr. and Mrs. Will Howes. Mr. Williams is engaged in the jewelry business at Paintsville, and is a model young man in every respect.

Those present were Mr. and Mrs. E. F. Howes, Mr. and Mrs. W. D. Huffman and Mr. and Mrs. Harry C. Howes.

***

The Paintsville Herald, May 11, 1916 — Rev. Partee, who is pastor of the First Baptist Church at Paintsville, preached for us at our church last Sunday evening. His subject was, "Systematic Giving, or Will a Man Rob God?" Rev. Partee is a man of advanced thinking and modern conception of Bible teaching. He was greatly enthused with his subject, and one could not help but see that he was filled with the spirit of it. His sermon was well received and did incalculable good.

There is no greater need in the church at this place, no more fruitful field for cultivation, nothing that will go so far to open our pocket books, and contribute to the awakening of the people of every interest of the church, than to occasionally have the privilege of hearing such sermons as this.

One of the first and most sacred obligations of any church, is the support of its pastor. No pastor can do his best work when the matter of the support of his family is constantly pressing upon his mind; when he is having to go in debt, borrow money,

Being a church deacon, sometimes Dad (left) was called upon to help baptize. James Lyons is on the right.

deprive himself of books and be grieved and humiliated because of the lack of encouragement along this line. Pastors as a rule do not expect large pay. They consecrate themselves when they enter the ministry, to a life of economy and frugality, but an adequate salary should be provided for and paid promptly. The average preacher is rather modest and reserved about mentioning this matter to his congregation, for he knows that if he says much, he is quite liable to be accused of being mercenary and after the dollar only.

There ought to be introduced into every church some sort of system for paying the pastor and every member, with rare exceptions ought to contribute something.

Clyde Roy Pack

Chapter 10

Tragedy: at Home and at War

This chapter contains six newspaper reports from *The Paintsville Herald* — chronologically listed — of Northeast Coal Company fatalities. As was stated earlier in this book, many editions of the paper that spanned the company's 47-year history are missing. Consequently, I feel certain this is not a complete list. As a matter of fact, two men who are not listed here were killed in 1922, their deaths resulting in murder trials for nearly two dozen union miners. Accounts of those trials appear later in the book.

Also reprinted here is an article about a Muddy Branch boy, Bruce Castle, who was killed in the service of our country.

***

The Paintsville Herald, September 25, 1924 — *Craig Castle, who for a number of years has been employed by the Northeast Coal Company at Thealka, was seriously injured last Saturday while working in one of the mines. He was removing a prop in one of the rooms and as the post was removed a large piece of slate was dislodged which struck him on the head. He was removed immediately to the Paintsville Hospital and his* condition is reported as serious, although he shows signs of improvement.

***

The Paintsville Herald, August 4, 1927 — Lacy Puckett, age about 30 years, was killed late last Thursday afternoon at Thealka, one mile west of Paintsville, when caught by a falling piece of slate which pulled loose from the roof of the mine in which he was working.

Mr. Puckett was a coal cutter and was operating a coal-cutting machine when the accident occurred. Two other men were working with Puckett at the time but managed to escape without injury. The piece of slate broke the neck of the unfortunate man and he died within a few minutes. It was five feet in length by four feet in width and 18 inches thick.

Mr. Puckett had been an employee of the Northeast Coal Company at Thealka for several years. He was a very popular young man and well liked by a great number of friends. He was a member of the Thealka baseball team and took an active part in all activities of the club.

He is a son of Mr. and Mrs. Will Puckett and besides his father and mother and several brothers, is survived by a wife and two children.

The body was laid to rest in the cemetery in Bridgford Addition, and was in charge of the Odd Fellow Lodge of Paintsville, of which he was an honored member.

***

The Paintsville Herald, August 18, 1927 — Roy Colvin, a miner employed by the Northeast Coal Company as a coal cutter, was painfully injured last Tuesday, and was brought to the Paintsville Hospital for treatment. An examination revealed that his leg was broken in three places. Colvin was engaged in cutting coal when the accident happened. His leg became entangled in the gear of the electric coal cutting machine and the bones snapped in three places.

***

The Paintsville Herald, April 26, 1928 — Bruce Preston, age 40, a resident of Thealka, and employed as a coal loader at No. 3 mine at Thealka, was instantly killed Tuesday afternoon when hit by a large piece of slate which fell on him. The slate struck the back of his head and he died instantly.

The remains were brought to the Paintsville Furniture Company undertaking department, and prepared for burial.

Preston was a native of this county and resided near Thealka for a number of years. He was a son of George (Bearhunter) Preston and was well known.

He is survived by a wife and five children.

Funeral services and burial were held Thursday.

***

The Paintsville Herald, January 26, 1939 — Bradley Stapleton, age 37, of Thealka, died at the Paintsville Clinic Wednesday. Mr. Stapleton was a miner for the Northeast Coal Company at the Thealka operation and was injured Monday of last week while operating a coal-cutting machine.

Funeral services will be held at the home Friday in charge of the Paintsville Furniture Company.

Mr. Stapleton is survived by his wife and six small children.

***

The Paintsville Herald, September 14, 1939 — Dow Castle, age 55, was killed Monday, September 11, in the mines of the Northeast Coal Company at Thealka. He was crushed to death by a slate fall in the room where he was working. Death, it is said, was instantaneous.

Mr. Castle had been an employee of the Northeast Coal Company since the company began operations at Thealka 30 *years ago. The company has had few fatalities at the Thealka operations and Mr. Castle's death is deeply deplored by the company officials as well as citizens of that community.*

Besides his widow, Mr. Castle is survived by eight children in addition to numerous other relatives and friends.

Funeral services were held at the home at 10 o'clock Wednesday morning. The services were in charge of Rev. Scott Castle and Rev. Millard VanHoose. Burial was made in the family cemetery on Road Branch in charge of the undertaking department of the Paintsville Furniture Company.

The Paintsville Herald, January 14, 1943 — *The War Department Saturday notified Mr. and Mrs. Matthew Caudill, Thealka, that their son Pvt. Francis M. Caudill, had been seriously wounded on the African battlefield. Caudill was shot on December 2.*

Known to all his Johnson County friends as Frank, Private Caudill joined the Army in 1941, when 20 years of age. He has been in foreign service the last seven months.

People of this community extend their sympathy to the parents and sincerely hope that encouraging news may soon be received concerning his condition.

Coal-Camp Chronicles

Mr. and Mrs. Caudill have another son, Pvt. Pierce Caudill, who left the states December 22 for foreign service.

Mr. Caudill is at present employed in New York.

***

The Paintsville Herald, July 23, 1943 — Last rites were held for Ammon Williams, 32-year-old victim of a mine slate fall on Saturday July 17, at the Free Will Baptist Church in Sitka. Burial followed in the Turner family cemetery at Sitka. Revs. Don Fraley and Fillmore Gambill officiated at the service. Pall bearers were members of the crew with whom he worked in the mines.

The accident occurred on Wednesday, July 14, in the Northeast mine at Thealka. Williams being caught under the slate fall. Death came on Friday at 2 a.m. Several other miners working nearby narrowly escaped injury.

Williams, a member of the Free Will Baptist Church, was the son of Frank and Linda Bush Williams of McDowell, Ky. He was born in this county on April 8, 1911. His wife, the former Lida Turner, and three children, Patricia Ann, Carl Thomas and James Roger, survive. Patricia Ann was ten years old the day on which her father was buried. He also has one brother, Ronald Williams.

D. Powell Williams, the late sheriff of Johnson County who was killed last year when struck by a train, was the young man's uncle.

The Jones Funeral Home had charge of arrangements.

***

The Paintsville Herald, February 17, 1944 — *Stephen Pack, 32-year-old miner of the Northeast Coal Company's mine at Thealka, died in a local hospital Tuesday afternoon from injuries suffered in a slate fall Monday afternoon. Funeral services will be held this afternoon (Thursday) at 2:30 at the church at Thealka. Burial will be made in the Thealka cemetery under the direction of the Jones Funeral Home.*

Mr. Pack had had 14 years of experience in the mines. According to officials of the mine, at the time of the accident, Pack was working in a 10-man crew which was loading the first cut off an entry where the top had been brushed. When the coal was undercut the immediate roof came between the pan line and the face when the rock broke off. Pack was pinned between the conveyor and rock.

Deceased is survived by his wife, the former Lizzie Hampton, daughter of Harry Hampton, and four small children.

Coal-Camp Chronicles

The Pack family has the sympathy of all who know them as the death angel has been a frequent visitor in their home in recent months. Both Stephen's mother and sister having passed on in that time.

***

The Paintsville Herald, May 10, 1945 — *It has been reported that Bruce Castle, husband of Irene Castle and son of Mr. and Mrs. Sam B. Castle of Thealka has been killed in action in Italy. He was 27 years old and was inducted on May 18, 1944. He was employed by the Consolidation Coal Company prior to entering military service.*

***

The Paintsville Herald, October 16, 1947 — *Floyd Jackson, city, and Ernest Robinson of Thealka, were injured today in a slate fall in the mine of the Northeast Coal Company at Thealka.*

Jackson, who is a patient in the Golden Rule Hospital, sustained bruises of the back and chest.

Robinson, a patient in the Paintsville Hospital, suffered a leg injury.

The men were not considered to be seriously injured.

***

Two of the aforementioned miners who perished in the Northeast Coal Company mines were my uncles. Stephen Pack, who was killed in 1944, was Dad's youngest brother. Bradley Stapleton, who died in 1939, was married to Dad's sister Zelphia.

Chapter 11

Northeast, According to the *Herald*

Time and careless stewardship seem to have destroyed official records of the Northeast Coal Company. However through news stories from The Paintsville Herald, one can derive a sense of the company's essence, its reason for being, the high and low points of its 47 years of operation, and the individuals who were responsible for its existence.

***

The Paintsville Herald, September 19, 1907 — The Northeast Coal Co. has engaged Gordon Burgess to teach a school for the benefit of the children of those engaged in its mines on Muddy Branch. The school is now in progress and the children are taking great interest in their studies. The school in no way interferes with the common school being taught on that branch. The children of those working for the Northeast Co. are so far away from the common school district's house, the company thought best to open a school and employ its own teacher.

This is the first instance in Eastern Kentucky where a mining company has manifested so much interest in the welfare of the children of those employed by it. Usually a mining company has no interest in those working for it. To say the least, money is rarely expended by a mining company in school and church affairs. The Northeast Coal Co., however, is an exception. A splendid school and church house has been erected near its mines. The building is large, well lighted and ventilated and provided with comfortable seats. A Sunday school is held in the building each Sunday morning and preaching Sunday morning and night. The Northeast has manifested the proper spirit and has set an example that would be well for other mining companies to follow.

***

The Paintsville Herald, November 28, 1907 — Mr. Smith, president of the Northeast Coal Co., operating on Muddy Branch, would like to change the name of that placid stream and has more than once criticized the Herald for referring to his coal operations as being on Muddy Branch. Mr. Smith is a whole-souled gentleman for whom we have the greatest respect, but to the Herald and the people of this section, Muddy Branch will ever remain Muddy Branch, it matters not what changes may be made.

Back under the old dispensation we used to do all our fighting on Muddy Branch, held our exciting school elections there, run

Coal-Camp Chronicles

our horses up and down that stream and take an occasional crack at random.

'Twas on Muddy Branch we held our barbecues the night before the election, and the occasion would draw some of the leading statesmen from Paintsville. No, a thousand times, no, we will always call her Muddy Branch. Our people have grown better and so has Muddy Branch, but so long as we live she will be known to us as Muddy Branch. The tearing down of the old

This old postcard shows the Northeast Coal Company store (left) and the boarding house during one of the frequent floods, probably in the late 1920s.

houses and the building of neat cottages, the construction of a railroad and whatnot, may in some respects change the appearance of things, but she is the same old Muddy Branch in a new dress.

***

Northeast's High Bridge opening on Teas Branch as it looked in the 1920s. Coal was hauled across the high bridge to the tipple at Muddy Branch.

The Paintsville Herald, March 29, 1917 —— Beginning April 1st, nine hours will constitute a day's work for employees of the Northeast Coal Company. In addition the miners will receive three cents more on the ton for loading the coal and all day laborers an increase in wages, and nine hours for a day instead

Coal-Camp Chronicles

of ten as heretofore. This is the second raise in wages the Northeast employees have received this year notwithstanding the fact that the company has not profited by the prevailing high prices of coal, having previously sold it on contract at a very close figure. But they are making this sacrifice in order that their employees may be able to meet the high cost of living, which, however is not so keenly felt by them as it is in many other sections.

On a great many articles carried in the store, special prices are made to employees, such as potatoes which are sold them for $1.60 per bushel. The same is true for many other staple articles such as feed, beans, corn, etc. They are furnished the finest of lump coal for only $1.50 per ton delivered. They are charged only $4.50 and $6 rent for 4-room and 5-room houses, which are lighted with electricity and for which no extra charge is made. These same houses, if in town, would rent for $15.00 per month.

Corporations are often referred to as "soulless" but this can not in justice be applied to the Northeast Coal Company.

***

The Paintsville Herald, November 1, 1917 — *Much praise is due Manager LaViers and the men who are responsible for the*

Northeast Coal company for the able manner in which they displayed their patriotism in helping their country in the Liberty Loan campaign. All share in this from the manager down to the coal miners, as all bought bonds.

The Northeast's high bridge as it looked from the east side.

Mr. LaViers was interested in the success of the Liberty Loan as he is always interested in anything that will help the town, state or nation. At Thealka $12,000 in bonds were sold.

The employees of the company are well pleased with the efforts to do their bit for their country. The company loaned the men money to buy bonds where they wanted it.

When it comes to doing the right thing at the right time, you can always put your finger on Manager LaViers and the companies he represents. They are deeply interested in our section and the welfare of our people.

***

The Paintsville Herald, January 1, 1918 — It is not only interesting but profitable sometimes, to take a retrospection of the past, not only of man's personal life but the life of his state, county, section. And in contrasting the present Big Sandy Valley with the Big Sandy of 10 years ago, one will see the greatest development and the most wonderful improvements on every hand.

For thousands of years the hidden wealth remained a secret, and as years rolled on and men in this century were struggling for supremacy in the world of commerce and finance, their minds naturally turned to this section where nature seemed to hold in store for mankind the wealth untold. Investigation led to discovery and discovery to development and the result today is that there is more coal mined in this valley than in any other like section of the state. Big Sandy is known far and wide for her rich and prolific coal beds and seams, and numerous operations are now in progress mining this product to be shipped to citizens in a less fortunate portion of the country, where coal fields are unknown.

Of the many mines that are being worked in the valley, none take precedence over the Northeast Coal Company. The main offices of this company are located in Paintsville, occupying the entire second floor of The Paintsville National Bank Building. These offices are the most modern and convenient in the valley.

At Thealka the operations consist of three large openings where a large number of men are employed at good wages and where local conditions are ideal. The men are well paid, well housed and given the very best of treatment in every way.

Thealka is a model mining town, made up entirely of local people, no colored or foreign labor being employed. A large store supplies the needs of the people with everything to eat and wear at the most reasonable prices. In fact the prices are lower than the miners can get the same goods for with cash at other stores. The miners are happy and prosperous.

Schools and churches are two things that go hand in hand with the operations of the Northeast Coal Company. The management figures that good schools and churches, with well-paid and contented miners are valuable assets to their towns and business.

At Thealka, the daily output is 1,000 tons from the three mines. This is also an excellent quality of coal.

Mr. A. Dw. Smith is president of the company and spends a considerable portion of his time in this section. Mr. Smith is well liked by all his employees and is ever ready to assist and help those working under him.

A. Dw. Smith

H. LaViers is general manager and is a citizen of Paintsville. He is an experienced coal man and under his able management the operations of this company have been successful and pleasing to both owners and employees. He is always willing to meet the employees of the company more than half way, and it is to his ability and liberality that all employees are satisfied with their conditions.

R. C. Thomas is superintendent of the operations at Thealka. He has had considerable experience in the coal business and is a favorite with not only his employees, but with all who know him. In addition to being a good coal operator, he is an ideal citizen and finds time from his many duties looking after the operations of his company to help out in all worthy causes. He resides in Paintsville and is a progressive, popular citizen.

R.C. Thomas

At Thealka the company operates a machine shop where all its work is done by the latest improved methods. Experienced men are in charge and in addition to their own work, a considerable amount of work for others is done. It is said to be the best machine shop in the Sandy Valley and is a valuable asset, not only to the company but the public as well.

No company in the entire valley is more interested in good roads and the advancement of this section than is the Northeast Coal Company.

Johnson County is indeed fortunate in having this operation within its borders, and its value to this county can not be overestimated by the people.

***

The Paintsville Herald, January 24, 1918 — The Northeast Coal Company has supplied the people of Paintsville and vicinity with coal during the winter at a price lower than they get for the coal f. o. b. cars at Thealka, in addition to paying for the delivery of the coal. They have supplied the people as a matter of accommodation and all our people have been supplied with coal.

When the fuel administration set the price to be charged for coal delivered locally, it was higher than the Northeast Coal

Company was charging for coal. If it had not been for the company, Paintsville would have been compelled to pay a much higher price.

The company is to be commended for this act and our people were certainly fortunate in having such a company located so near the town. Not a single request for coal was refused by the company.

The Herald feels that a word of praise for this company is not out of order at this time.

***

The Paintsville Herald, February 1, 1923 *— Jess Fletcher, one of the men charged with the explosion in Thealka mines last year, is now on trial here. He is one of 19 men indicted, most all of whom have confessed to the part they had in the explosion. Fletcher has not made a confession and claims his innocence, but the Commonwealth claims that the men who have confessed implicated him in the affair.*

He is being represented by Attorney Blaine Clark, of Inez, and W. J. Ward of this city, while the Commonwealth is being assisted by Attorney M. C. Kirk of this city and Attorney Jack May of Prestonsburg.

A jury from Lawrence County is hearing the evidence and the Commonwealth closed its testimony Wednesday and the case was ready to be submitted to the jury as The Herald goes to press.

The other men indicted with Fletcher will be tried at a special term of the court, probably April 16.

***

The Paintsville Herald, July 19, 1923 — *Trial of the men charged with the explosion that killed Cove Smith and Will Helton, non-union miners at Thealka mines near Paintsville, started in the Johnson Circuit Court Tuesday after a jury from Floyd County had been selected. Seventy-five Floyd County citizens were summoned here to select the jury from. Monday was taken up with the selection of the jury and Tuesday afternoon the jury was completed, composed of the following: J. P. Sturgill, P T. Cline, Ed Baldridge, Roe Hyden, Dave Gobel, G. G. Prevett, J. H. Wells, Harry Prevett, Jake Setser, W. R. Wells, W. T. Prevett and John W. Gobel.*

The case is the result of the miners strike last year in which union miners went out on strike and the Northeast Coal Company operated their mines with non-union labor. The two men killed were from Morgan County and went in the mines on the day of the explosion to operate machines which it is claimed were charged with explosives by union miners. They were the only two men to reach the machines when the explosion occurred. Both were instantly killed.

Detectives were placed on the case and in a few weeks evidence was secured which resulted in the arrest of a few of those accused and a confession of a number of them put the required information in the hands of the authorities, causing the wholesale arrests.

The first man to be tried is Jess Fletcher. It will require several days to complete the case. It is not thought that all of those accused will be tried at this term of court.

Much interest is being shown in these cases and we hope to be in a position to give full details in the next issue of The Herald.

Indicted with Jess Fletcher in this case are Ed Pelphrey, Frank Conley, Bill VanHoose, Josephus VanHoose, Ed VanHoose, Dave Castle, Ed Moore, Everett Moore and Burns Castle. Other indictments include the names of a large number of others who will also be brought before the court at this term.

***

The Paintsville Herald, July 26, 1923 — *Jess Fletcher, first of the defendants in the Thealka mine explosion case was found* not guilty of the charge of murder of Cove Smith, by a Floyd County jury, in the Johnson Circuit Court Thursday afternoon. The case had been before the court for four days, three days being taken up by the hearing of evidence from both sides. The jury took the case last Thursday afternoon and after one and one half hours deliberation, returned the verdict of not guilty.

This is the first of a number of trials that were called by this court as a result of an explosion at the No. 3 mine of the Northeast Coal Company at Thealka, June 27, 1922. Two non-union miners were killed by an explosion thought to be dynamite which the evidence shows had been placed in a coal cutting machine and had been so wired with electricity as to explode the charge when the current was turned on in the machine.

A number of arrests were made and several of the men arrested, confessed to the obtaining of the dynamite from the storehouse of the Northeast Coal Co., but the prosecution failed to show that Fletcher was connected with the obtaining or the placing of the explosive.

The testimony of some of the witnesses said that there had been a meeting of the union miners in the Stafford Bottom across Paint Creek from Paintsville in which they planned to procure the dynamite and place it in some part of the mine where it would explode and scare the non-union miners away.

This explosion occurred while the union miners were out on a strike and the company was running the mines with non-union labor.

The other men will be tried by a jury from Boyd County which came to Paintsville Monday. Seventy-five citizens were summoned from which the jury will be chosen.

***

The Paintsville Herald, July 26, 1923 — *A judge's ruling on a motion for a peremptory dismissal of charges against the defendants today looms as the feature of the trial of six Johnson County miners here in which the evidence relative to the explosion in one of the Thealka mines of the Northeast Coal Company last year, resulting in the death of two men, is being heard for the third time.*

The motion was made late yesterday after the Commonwealth rested its case but Judge J. Frank Bailey reserved his ruling until all the evidence has been heard. This is expected to be completed late today and the court's decision will follow immediately.

The motion for peremptory instructions is based on the argument of the defense that the state has introduced no evidence in this trial to corroborate the testimony of the three accomplices. Counsel for the defense declared in explaining its motion that the state had failed to connect the six defendants with the explosion except by testimony of the three accomplices which must be corroborated before it can be competent.

Opposing counsel presented the claim that the testimony of H. LaViers, general manager of the Northeast Coal Company, relative to the condition of the roof of the magazine from which the dynamite used in the explosion is alleged to have been taken, corroborates the evidence of the three accomplices because their stories of how the magazine was entered tallies with the manager's description of the conditions.

Counsel for the defense refuses to grant that this is corroborative testimony and as the Commonwealth made no attempt to show the court whereby it had presented other corroborative testimony, the Judge is compelled to rule on the motion with little to aid him in making a decision.

The six men being tried are Drewery Castle, Albert Green, Harrison Castle, Russell Ramey, Smith Conley and Grant Fletcher. They are charged with conspiracy and murder in connection with the deaths of Cove Smith and Will Helton, two non-union miners in the explosion in the Northeast Coal Company's No. 3 mine at Thealka on June 27, 1922.

This is their first trial, although the evidence being used was heard in the two trials of Jess Fletcher, acquitted of murder in court here last week.

The State has used three witnesses, Norton Rice, Will Spears and Ed Pelphrey, indicted on the same count as the six defendants, who have testified that they were present at meetings when they and the six defendants planned the explosion and when some of the defendants stole the 100 or more sticks of dynamite from the company's magazine.

These three witnesses said that the roof of the magazine was broken into on the Saturday night preceding the explosion and the dynamite was taken. Though the state has introduced no testimony to connect the defendants with the theft of the dynamite or the ensuing explosion, it did present the evidence of the company manager who said that he had examined the magazine months after it had been broken into and had found where the roof had been pried up and bent back.

The fate of the six defendants therefore hangs upon the imprints of a crowbar on the roof of the magazine. The court is required to rule whether this is or is not corroborative testimony. If he rules that it is, the case will go to the jury, but if he rules that it is not and the Commonwealth is unable to show whereby it has presented other corroborative testimony, he will be forced to issue instructions for the dismissal of the defendants.

The defense has also moved to exclude the testimony of the three accomplices on the ground that it has not been corroborated. It also objected to the testimony of Henry LaViers relative to his description of the condition of the magazine. The court reserved ruling on these questions until all the evidence is heard.

The defendants denied the stories told by their three alleged accomplices, denied attending the meetings mentioned, denied stealing the dynamite, denied the existence of a conspiracy and denied placing the dynamite in the coal cutting machine.

The remainder of their testimony consisted chiefly in their showing their alleged activities and whereabouts on the days and nights mentioned by the Commonwealth's witnesses. The other witnesses were used to attempt to strengthen the alibi of Smith Conley.

On rebuttal both sides will attempt to ridicule the stories of opposing witnesses, with the state trying hard to break the alibis of the defendants.

Attorneys for the state are Commonwealth Attorney John W. Wheeler and County Attorney Sam Stapleton, A. J. May of Prestonsburg, H. S. Howes, M. O. Wheeler and M. C. Kirk of Paintsville.

Attorneys for the defense are C. B. Wheeler of Prestonsburg; J. B. Clark of Inez; W. J. Ward, I. G. Rice, and Blair & Harrington of Paintsville.

***

The Paintsville Herald, August 2, 1923 — *The six defendants on trial last week in the Johnson Circuit Court on charges of murder and conspiracy in connection with the deaths of two men in a dynamite explosion in a mine at Thealka near here last summer, were acquitted by a Boyd County jury in a verdict at 10:55 Thursday morning.*

The jury was given the case at 10:30 Wednesday night after a trial lasting three days and nights. It began its deliberations at 6 a.m. and brought in a verdict four hours and fifty-five minutes later. It is understood the first ballot, taken shortly after deliberation was started, stood at ten for acquittal and two for conviction.

The six men freed by the verdict are Drewery Castle, Albert Green, Harrison Castle, Smith Conley, Grant Fletcher and Russell Ramey.

The case was given to the jury at 10:30 following the arguments which began at 7 o'clock. The evidence was completed and court was adjourned at 5 o'clock. Two arguments were made before the jury. The two of them required three hours and a half, a remarkably short time considering the seriousness of the charges and the number of the defendants. The argument for the defense was made by Judge C. B. Wheeler of Prestonsburg. He spoke first and was followed by A. J. May of Prestonsburg who made the only argument for the state.

As the case drew to a close, interest seemed to be lagging. The defense occupied all of the morning session and until three o'clock in the afternoon with its evidence. This consisted chiefly of attempts to strengthen the alibis of the six defendants and to discredit and refute the witnesses for the state. There was only one break in the monotonous routine of evidence and that was the appearance H. B. Emmerson, a St. Louis detective, employed by the Northeast Coal Company to unravel the *mystery of the explosion, on the stand. His testimony consisted of denials of offering money to miners for information relative to the explosion but it was the man and not the evidence with which the spectators and the witnesses seemed interested. Much has been heard of this detective since the explosion last year but few of those interested in the case knew him. Interest also centered about the court's ruling on the motion of the defense for peremptory instructions, a motion to exclude the testimony of the three alleged accomplices, and objection by the defense to the testimony of Henry LaViers, general manager of the Northeast Coal company, relative to the condition of the roof to the company's magazine from which the dynamite was said to have been stolen.*

All three matters were overruled by the court. The court held that the condition of the magazine was a corroboration of the testimony of the three accomplices, and therefore it was sufficient evidence to submit the case to the jury. The hinging of the case on the ruling of the court on this motion for a peremptory dismissal caused the fate of the six defendants to rest on the imprints of a crowbar on a piece of roofing. Norton Rice, Will Spears and Ed Pelphrey, the three alleged accomplices under indictment for the same count with the defendants, testified that the magazine was broken into on the Saturday night preceding the explosion.

They testified that entrance was gained after a portion of the metal roofing had been pried up and bent back. The testimony of Henry LaViers to the effect that he found the magazine in that condition when he examined it several months following the explosion tended to substantiate the three accomplices, according to the commonwealth.

***

Copies of *The Paintsville Herald* were not available for the rest of the month of August 1923, and newspaper coverage of the trial, at least as far as we are concerned, ends here.

However, no one was convicted. From a trip to the Johnson County Circuit Court Clerk's office we learned that in *Criminal Order Book No. 3*, pages 348-350, several entries regarding the trials read as follows: "The Commonwealth, having investigated the foregoing case against [here the following names were listed on various pages] charged with wilful murder and believing after said investigation that evidence sufficient to convict said defendants cannot be had — on motion of Commonwealth, it is ordered that this cause be and the same is hereby dismissed and stricken from the docket. John W. Wheeler, Commonwealth Attorney.

The following names were listed: Ed Pelphrey, Frank Conley, Bill VanHoose, Josephus VanHoose, Ed VanHoose, Jess Fletcher, Dow Castle, Ed Moore, Grant Fletcher, Will Spears, Smith Conley, Harrison Castle, Drewery Castle, Russell Ramey, Albert Green, Norton Rice, and Everett Moore.

***

The Paintsville Herald, September 27, 1923 — *An unconditional denial, accompanied by affidavits, that Samuel Pascoe, president of District No. 30, United Mine Workers of America with headquarters at Ashland was beaten up by a mine guard of the Northeast Coal Company of Thealka on August 3, has been filed before the United States Coal Commission at Washington by the Bituminous Operators' Special Committee, it was learned.*

The statement followed one filed by the mine workers' union several weeks ago saying in substance that Pascoe had been beaten and continuously menaced in Paintsville. Pascoe alleged that his jaw had been broken in the fight and that hospital treatment had been refused him at the Paintsville Hospital.

Pascoe told the Independent *three weeks later that he had been forced to go to Indianapolis, Ind., for treatment and that he had remained for ten days in the hospital there. He also said to the* Independent *that he had been refused medical aid at Paintsville.*

The statement in which the affidavits are summed up, was prepared by Col. Henry L. Stimson and Goldwaithe H. Dorr, counsel for the Bituminous Operators Special Committee. It, with the letter accompanying it, follows:

•

September 7, 1923
U. S. Coal Commission
Washington, D.C.
Gentlemen:
On August 28, 1923, the International organization of the United Mine Workers filed with your Honorable Commission a statement as to an assault upon one Samuel Pascoe. This is the first specific charge of an alleged encroachment upon the civil rights of one of its members which has been brought to our attention and to which we have had any opportunity to make a reply. The charge is that Pascoe, President of Dist. No. 30 of

the Organization was severely beaten at Paintsville, Ky., on August 3, 1923, by a hired guard of a coal company; that Pascoe went to a hospital at Paintsville for first aid treatment but was refused it, and the inference is drawn that this refusal to give him treatment was on account of the fear of the hospital authorities that they would incur the displeasure of the coal companies.

We have investigated this incident and wish to present the true facts to your commission.

Samuel Pascoe is the man who led the 1922 strike in the course of which two men were killed near Paintsville and others wounded and shot at. He publicly made a speech near Paintsville which was heard by many people and currently reported, in which he stated, "We will win this strike if we have to wade knee deep in blood to do it."

In another speech made near Paintsville, he urged the striking miners "to use their 45's if necessary to prevent the mining of coal." He incurred the enmity of many miners who tried to work during the strike. In continuing the strike after it was hopeless, he also incurred the enmity of many who had formerly supported him. By the violence of his language and his leadership of a reckless and wholly unjustified strike, he thus had made himself a veritable storm center in the community of Paintsville.

The man who assaulted Pascoe on August 3, 1923, was not a mine guard. The only coal company at Paintsville employs no mine guards whatever. During the strike in the summer of 1922, after repeated acts of violence had been committed by the Mine Workers, this man had been employed to guard the property of the coal company near Paintsville and to furnish protection to men who desired to work there. His employment as a guard ceased with the end of the strike in the early autumn of 1922.

He had been present at a meeting where Pascoe made insulting remarks about men who continued in the employment of companies during the nationwide strike, calling them "lower than the slums of the world," and said that their wives should leave them and that no decent woman ought to live with them. He had taken offense at these remarks and resolved to avenge himself on Pascoe whenever the opportunity presented itself. The day of the assault was the result of his personal animosity against Pascoe.

Finally, when Pascoe went to the only hospital in Paintsville to secure treatment, he was immediately admitted and given prompt and efficient attention and treatment. The hospital *authorities offered him a room in which he could stay until he had more fully recovered. He refused this offer. The hospital is largely supported by the coal companies whose experience with Pascoe's followers in 1922 is described on pages 12 to 14 of our brief No. 3 on Northeastern Kentucky.*

We are prepared to support this statement by legal proof. A copy of this letter has been sent to the United Mine Workers.

Respectfully,

Henry L. Stimson

Goldwaithe H. Dorr

Of counsel for the Bituminous Operators' Special Committee

•

A number of affidavits on file in Washington disprove the formal communication of the United Mine Workers of America to the United States Coal Commission to the effect that Samuel Pascoe, District President of District 30 Union agitator in Northeastern Kentucky, was assaulted recently by a mine guard, with the inference plain that the assault was ordered by coal operators.

Link Castle, the man who attacked Pascoe, has sworn to an affidavit that his attack was the outcome of a purely personal grudge, growing out of an insulting speech Pascoe made, attacking the men who worked at Thealka, Ky., during the strike last year. Castle is not a mine guard.

"It has been rumored that the Northeast Coal Company hired me to do this," says Castle's affidavit, speaking of the affray, "but this is absolutely a falsehood. The Northeast Coal Company didn't know anything about this until after it happened and they had in no way, shape, form or fashion anything to do with it. This is only a personal matter between Pascoe and myself."

Affidavits of three doctors flatly refute the contention of the union that Pascoe was refused treatment when he claimed that his jaw was broken. The doctors swore that Pascoe came to the hospital and was thoroughly examined. They found the patient had suffered a black eye, another bruise on the face, and a bruised and swollen jaw. There was no dislocation, the patient talking normally and having full use of his jaw, but as he, Pascoe, feared that a fracture existed, it was examined by the fluoroscope and found to be in tact. Pascoe was advised that there was no fracture, but that it would take several days for the discoloration to subside, and that if he desired to remain at the hospital for further treatment he could have a room. Pascoe *declined, saying it was necessary for him to return to Ashland that evening.*

One affidavit reviews the attacks during the 1922 strike of workmen at Thealka by strikers who cut the power house lines on one occasion, and three times shot at working miners from ambush. It was not until after the third shooting that deputy constables were appointed as guards to protect the men remaining at work. Castle, for a time, was one of these constables who was employed only during the period of the strike of 1922, and with the ending of the strike in the month of September, 1922, they were relieved from their duties with the company as employees producing coal. There have been no guards used since that time nor were there at any time guards used previous to the occurrences in the year 1922, referred to herein.

It was while Castle was a mine guard, working under the direction of a constable elected by the people, that Pascoe made the statement in a public speech that no decent woman would live with a mine guard, and similar insulting remarks. In the same speech, according to the affidavits, Pascoe referred to the late President Harding as "an old man shaking with the palsy before John L. Lewis, President of the United Mine Workers of America, when they were negotiating the strike settlement."

Castle swore himself that he had made up his mind to avenge himself on Pascoe, and that the first intimation he had of Pascoe's presence in Paintsville was when he saw him sitting on the porch in front of a restaurant. Whereupon he, in the words of his sworn statement, "was as good as my word."

It was only during the period of this strike that guards of any kind have ever been used in the district, and their use at the time was forced and necessary for the full preservation of law and order. Previous to this time, nor since the ending of this strike, there have been no guards of any kind on duty anywhere in the entire Big Sandy district of Eastern Kentucky.

***

The Paintsville Herald, September 25, 1924 — The Northeast Coal Company of Thealka, Auxier and White House, has been having good runs with their several mines. We are also informed that the prospects are good for steady work in the future.

This will come as good news to the people of this section as the Northeast Coal Company only employs native labor and the money paid for labor is spent in our own country.

It is possibly the only large operation in the valley that employs native labor exclusively.

***

The Paintsville Herald, September 12, 1929 *— The Northeast Coal Company First Aid team, namely, Johnny Preston, Delmas Preston, W. M. Wallen, Ora D. Spears, Russell Conley and V. D. Picklesimer, team captain, winners of the cup in the state meet at Lexington, August 31, left Monday morning by motor for Kansas City, Mo., to represent Kentucky in an international meeting. The Northeast boys came out with a score of 598 points out of a possible 600 points. This was an excellent score and there were teams just a point or so behind, which goes to show that there are others working toward the same goal.*

While our boys are not natives of the "show me" state and know that they may not win the cup this time, they are, nevertheless, believers in Missouri's philosophy.

This meeting is open to all mines and other industries that wish to compete. It is held annually and in 1931 the meeting will be held in Lexington.

Coal-Camp Chronicles

The results of the contest will not be available until Saturday afternoon. Keep your eye on Kansas City and boost for your hometown team.

Another Northeast Coal Company first-aid team, probably taken in the late 1940s or early 1950s. Ernest Dove is standing (center), back row. Archie VanHoose is seated (second from left) and Charlie Bailey is seated (second from right). Others pictured are Bernard Castle, Elmer Castle, Grant Davis and Richard Castle.

Clyde Roy Pack

***

The Paintsville Herald, September 19, 1929 — *Kentucky's Champion First Aid Team, residing at Thealka, near Paintsville, was the guest of the Paintsville Rotary Club last Tuesday at the regular noon-day luncheon held in the Mayo Memorial Church. These six popular young men were the guests of Mr. Henry LaViers, general manager of the Northeast Coal Company. Mr. Laviers, in a splendid address, introduced the young men and gave them a boost and described their work.*

Virgil D. Picklesimer, captain of the team, made a splendid talk on his team's activities and their duties. He told the club how the team won the state championship, and how they were up among the best teams in the United States in the international meeting held at Kansas City, Missouri, last week.

***

The Paintsville Herald, April 4, 1940 — *Fred L. Sherman, who has been with the Northeast Coal Co. for the past 27 years, succeeds R. C. Thomas as superintendent of the company's operations at Thealka. Mr. Sherman assumed his new duties Monday morning.*

Mr. Sherman has been mine foreman at the Auxier operations for the past nine years and has been well liked by his fellow workers. *He is an experienced coal man. He is efficient in his work and has taken every interest in the company and the welfare of his men.*

He began his mining career at Thealka and has worked his way from the bottom and deserves the promotion which has been given him in recognition of his efficient and loyal service to the organization. He has many friends in the Big Sandy Valley who wish him success in his new position. Mr. Sherman is a native of Johnson County and was reared at Paintsville.

***

The Paintsville Herald, January 31, 1935 — *Morris Williams, 79, retired coal operator and nationally-known authority on the coal mining industry died at his home in Philadelphia last Friday. Mr. Williams was apparently in excellent health and his death came as a shock to relatives and business associates.*

Mr. Williams had been identified with the coal industry in Eastern Kentucky for years. He was one of the founders and principal stockholders in the Northeast Coal Company and Glogora Coal Company and had been one of the leading figures in the development of the industry in Kentucky, West Virginia and Pennsylvania.

He was born in Wales and came to the United States at an early age with his parents. He entered the mining business in early manhood and at one time operated a gold mine in the West. He rose from an obscure place in the industry to one of the leading figures.

He first became interested in the Big Sandy Coal field more than 30 years ago when he and associates organized the Northeast Coal Company here. Later he and associates organized the South East Coal Company at Seco on the Kentucky River. In 1920 he organized the Glogora at Glo.

***

The Paintsville Herald, February 10, 1944 — *A. Dw. Smith, a prominent coal operator of Eastern Kentucky died at his home in Philadelphia Monday morning on February 7. Mr. Smith was eighty years old. He came to Big Sandy in 1905 with the late Morris Williams, and purchased the old Keyser Coal Company from the late Charles M. Keyser and John C. C. Mayo. This was the beginning of the Northeast Coal Company. Later they opened mines at Auxier and only a few years later built the towns of Seco and Millstone and opened mines of the South East Coal Company of that place.*

Coal-Camp Chronicles

Associated with this extensive development was Mr. Henry LaViers of this city, and under Mr. LaViers' management the mining towns won the reputation of being among the cleanest and most progressive mining towns in Eastern Kentucky.

Until recent years, Mr. Smith visited Paintsville and Eastern Kentucky every month. He took a deep interest in the civic and educational welfare of the community and contributed largely from his means for the betterment of conditions in the town and county.

He was an honorary member of the Paintsville Rotary Club and that organization at its meeting Tuesday adopted resolutions of regret and sympathy.

Many will remember Mr. Smith as the kindly, genial gentleman that he was, whose influence was always on the side of every good movement that had as its purpose the upbuilding and progress of the community. To his many friends hereabout, his passing away is a matter of sincere regret. Let us who remain to carry on be imbued with the same lofty spirit and his ideals that characterized him in our community life and interest.

Mr. Smith is survived by one daughter, Mrs. Alden Taft of Philadelphia, Pennsylvania, and one son, Mr. Alan Smith of

Cincinnati, Ohio, who succeeded his father as president of the South East Coal Company.

***

The Paintsville Herald, April 4, 1946 — Richard Collins Thomas, 69, General Manager of the Northeast Coal Company, died suddenly at his home here of a heart attack on Monday, only seven weeks after the death of Mrs. Thomas. He had been ill for a few days but his condition was thought to be improving. He suffered the attack about 9:36 a.m. while sitting in his chair, and he died before medical aid could be summoned.

Mr. Thomas was born in Maryland on May 26, 1877.

He was the first mayor of Paintsville and was a charter member of the Paintsville Rotary Club. He served in various capacities in community and civic organizations. His kindliness and generosity caused him to be loved by everyone who knew him and especially by the numerous employees at the Northeast Coal Company, many of whom had worked under his supervision for many years.

A member of the First Baptist Church, he worked diligently for the church and the uplift of the community. No citizen was held in higher esteem and he will be greatly missed throughout the Sandy Valley.

Surviving are seven children, Mrs. Francis W. Clay, Bowling Green; Mrs. Earl King Sneff, Morehead; Harry Thomas, city; Wildan Thomas, McPherson, Kan.; Mrs. John Shuey, LaFollett, Tenn.; Mrs. Jack Hill, city; and Mrs. Charles Morris, Frankfort. Also two brothers, David C. Thomas and Charles C. Thomas, Wellston, Ohio; two sisters, Mrs. Harry Harrell and Mrs. E. J. Harper, Wellston, Ohio.

***

The Paintsville Herald, April 1, 1948 — *A fire of undetermined origin destroyed the tipple of the Northeast Coal Company at Thealka early Sunday morning. The blaze was discovered by a passerby about 2 a.m. but the residents were unable to stop the flame, the town having no fire department.*

Three Chesapeake and Ohio railway cars were also destroyed in the fire.

The damage was estimated to be heavy and it was not learned whether the tipple will be replaced or when the employees may expect to return to work after the settlement of the nation-wide shut down.

The mine produced about 700 tons of coal a day when the plant was in operation. Approximately 200 men were employed and about 100 lived in the town. The tipple was rebuilt about 10 years ago and had been modernized and was equipped to grade the coal into five sizes.

A reward of $2,000 has been offered by the company for information leading to the apprehension, arrest and conviction of the person, or persons, responsible for the fire.

***

The Paintsville Herald, October 14, 1948 — *A modern coal tipple with a 155-ton per hour capacity has recently been completed by the Northeast Coal Company at Thealka, it was announced by Fred Sherman, superintendent of the company's mines.*

The new tipple has three tracks and will handle four or five grades of coal. Started in November of last year, it was built chiefly for the truck mines of the company and other truck mines in the locality. From 200 to 250 additional miners have been employed in the independent truck mines, and from 12 to 15 have been added to the company's payrolls since the installation of the tipple, Mr. Sherman said.

An additional screen is being installed this week.

A second modern tipple is being built by the North East Coal Company at its No. 3 mine at Thealka. This structure will replace the one burned on March 28, and will handle only the coal from No. 3 mine. Started on April 6, the tipple has four tracks and will handle 200 tons of coal an hour.

A new tipple was built in 1948.

Construction of a large storage bin with a 300-ton capacity is being built by the Hager Hill Coal Company at its mine near Sitka. The bin will be used for run-of-mine coal.

***

The Paintsville Herald, January 31, 1952 — The last remaining mine of the Northeast Coal Company at Thealka ceased operation Jan. 26, ending the company's 47 years in coal production in Johnson County.

The No. 5 mine of the company was closed about two months ago, and the closing of the No. 3 mine last Saturday necessitated the laying off of about 60 miners, it was announced by Fred Sherman, superintendent. The two mines when working at capacity produced about 300 tons of coal per day, he said.

The company will continue the operation of two tipples at Thealka for coal produced by surrounding truck mines. All other machinery, and steel, will be removed from the premises.

About 85 houses at Thealka have been sold to individuals during the past few years and about 17 five-room houses are yet to be sold, Mr. Sherman stated this week.

He added that the closing of the Thealka mines was caused by the poor quality of the coal which made it unmarketable.

Coal mining was started by the Northeast at Thealka in 1905.

***

I've no idea when Time suddenly began to pick up speed, but it certainly has. As remembered by the man the boy became, for this coal-camp kid, Time was a terrapin creeping through the weeds to eat ripe tomatoes in Dad's garden; it was the long walk home after darkness had driven us off the baseball field in the school house bottom; it was the week of worry and anticipation between the cliff-hanging episodes of the serial running at the Saturday matinees at the Sipp and Royal theaters.

But sometime when I wasn't paying attention, Time, like those transformer toys, became a gazelle. Since I've never seen one, at least it became what I've always imagined a gazelle to be. Time now makes the fleet-footed Mercury of Greek mythology look like an octogenarian in a 100-yard dash.

There was once a time in my life when I considered Time to be my buddy. When Mom would say, "Wait 'til your daddy gets home," I knew Time would go to work on her memory and make her forget whatever I'd done that she felt was too serious for her to handle herself.

When Mr. Chandler, our principal at the Muddy Branch school, would announce that the school nurses would "be here next Tuesday" to give us our yearly round of vaccinations, I knew that "next Tuesday" was light years away and lost no sleep over it ... at least not until "next Monday" night.

When distant radio stations began playing Christmas carols and giving nightly countdowns of the days left until Christmas; 30 ... 25 ... 12 days, it didn't mean enough for me to change my behavior.

Now it seems that Time has become a worthy adversary; and one that doesn't fight fair. When I wasn't looking, it sneaked up and pulled out a lot of my hair, making my forehead at least two inches longer. What it didn't pull out, it turned gray, at least around my temples. It caught me asleep and painted bags under my eyes. It implanted an inflated innertube just under my skin where my waist used to be.

Time cheats. It even fights dirty. But more than anything else, and as far as I'm concerned, the worst thing it has done is speed up.

When I was a kid, birthdays, like Christmases, came eons apart. When I was ten, I always said I was ten, going on 11. When I was 20, I couldn't wait to turn 21. But just as soon as I turned 21, I was suddenly 30. Then 40. Then 50. Then 60, and on and on.

If Time were a real person, I don't think I'd like him very much.

About the Author

Clyde Roy Pack is an associate editor at <u>The Paintsville Herald</u>, where he also writes an award-winning humor column. He was an elementary and high school art and English teacher for 33 years before retiring in 1994.

He is also the author of <u>Muddy Branch: Memories of an Eastern Kentucky Coal Camp</u>, published in 2002 by the Jesse Stuart Foundation (ISBN 1-931672-10-5).

He lives in Paintsville with his wife of 42 years, Wilma Jean Penix Pack.

Index

A
Adkins, Irene, 84
Ardigo, J.J., 53
Arrowood, Lillian, 133
Auxier, Bryan, 3

B
Bailey, Billy Boy, 21
Bailey, Charlie, 66, 80,183,185,234
Bailey, Frank, 217
Baldridge, Ed, 214
Baldridge, Liss, 143
Baron, George, 121
Beane, Ed, 53
Bellomy, M/M Abray, 75
Blair, Fat, 104
Brooks, Alma, 134
Brooks, Alpha, 134
Brooks, Bertha, 134
Brooks, Carmen, 134
Burgess, Frank, 80
Burgess, Gordon, 202
Burton, Bo, 137
Burton, Donna, 20
Burton, M/M E. L., 71, 72
Burton, Emogene, 137
Burton, Foster, 41, 83, 179
Burton, Mrs. Foster, 138
Burton, Irene, 84
Burton, John, 20, 134
Burton, Katie, 20
Burton, Mary Christine, 72,134
Burton, Mollie, 71
Burton, Pauline, 84
Burton, M/M R. C., 72, 74
Bush, Sophia, 80
Butcher, George W., 125
Butler, Claudia, 79
Butler, Elizabeth, 79

C
Cady, Sam, 53
Carder, Rev. O.J., 189
Castle, Alma, 80
Castle, Andrew, 86
Castle, M/M Beecher, 86
Castle, Bernard, 234
Castle, Billy Ray, 85
Castle, Bruce, 90, 134, 193, 200
Castle, Burns, 215
Castle, Carmel, 134
Castle, Curtis, 90, 134
Castle, Craig, 193
Castle, Darlene, 92
Castle, Dave, 215
Castle, Dow, 196, 225
Castle, Douglas, 88
Castle, Drewery, 218, 222, 225

Coal-Camp Chronicles

Castle, Elizabeth Mae, 92
Castle, Elizabeth, 80
Castle, Elmer, 234
Castle, Ernest, 90, 134
Castel, Esteel B., 92
Castle, Fannie, 87
Castle, M/M Frank, 86
Castle, Gailord, 80
Castle, Gene, 88, 89
Castle, Golda Mae, 134
Castle, Harrison, 218, 222, 225
Castle, M/M Herbert, 87
Castle, Ira, 134
Castle, Irene, 134, 200
Castle, Ivel, 80
Castle, James Roger, 21
Castle, Jessie, 87
Castle, Joe, 92
Castle, Johnnie, 134
Castle, Junior, 134
Castle, Kathryn 134
Castle, Link, 229
Castle, Lonnie, 54
Castle, Mrs. Luther, 88, 89
Castle, Mabel, 80
Castle, Mildred, 86
Castle, Minnie, 90
Castle, M/M Monroe, 85
Castle, Orville, 134

Castle, Prudence, 138
Castle, Richard, 234
Castle, Sam B., 90
Castle, M/M Sam B., 200
Castle, Rev. Scott, 197
Castle, Rice, 78
Castle, Ruie, 80
Castle, Ruth Aileen, 138
Castle, Sylvia, 86, 87
Castle, Tom, 21
Castle, Woodrow, 87
Castle, M/M Woodrow, 86, 88
Caudill, Amanda, 134
Caudill, Anna Mae, 134
Caudill, Francis M., 197
Caudill, Rev. L. F., 188
Caudill, M/M Matthew, 197
Caudill, Pierce, 198
Chandler, Avery Oakley, 134
Chandler, Garfield, 67
Chandler, Pauline Mae, 134
Childers, Loula, 85
Church, Clara, 87
Clark, Blaine, 213
Clark, J. B., 221
Clatworthy, Mrs. John, 74
Clay, Mrs. Francis W., 240
Cline, P. T., 214
Collins, Gladys, 134

Collins, Rev. T. J., 70, 71, 72, 73, 75, 76, 188
Colvin, Lizzie, 21, 86, 137, 145
Colvin, Lois Ann, 21
Colvin, Pick, 33
Colvin, Roy, 195
Colvin, M/M Roy, 82
Compton, Mrs. Lou, 79
Conley, Betty Helen, 88
Conley, Rev. Burns, 76
Conley, Clyde, 180
Conley, Charles Odes, 85
Conley, Mrs. F. J., 76
Conley, Flem, 74
Conley, Frank, 215, 225
Conley, Mildred, 134
Conley, Ray, 138
Conley, Rusha, 80
Conley, M/M Russell, 81
Conley, Smith, 218, 220, 222, 225
Conley, Solon, 80
Conley, M/M Stelson, 85
Conley, Stinson, 80
Corder, Nelle, 133
Corder, Mrs. J. E., 138
Cotton, Granny, 143, 177
Crichton, Michael, 152
Crider, Alfred, 86
Crider, M/M Alfred, 88
Curiutte, Esther, 134

D

Dale, Helen I., 89
Dale, M/M James, 89
Dale, Katherine, 136
Dale, Rev. Raymond, 183
Daniel, Mrs. Dennis, 136
Daniel, Eugene, 136
Daniel, Frank, 80
Daniel, Garner, 134
Daniel, Mona, 84
Daniel, Tive, 80
Daniel, Tucker, 21
Daniels, Carolyn Lyons, 3
Daniels, Jerry, 36
Daniels, Jim, 47
Daniels, John, 21
Daniels, P-Jack, 21
Daniels, Tiny, 21, 48
Davidson, Eugene, 80
Davis, Billy Boy, 134
Davis, Carroll, 134
Davis, Earl, 134
Davis, Edna Mae, 134
Davis, Ernest James, 134
Davis, Grant, 234
Davis, Helen, 134
Davis, Marcus, 134
Davis, Mrs. Willard, 81

Davis, Wilma, 134
Davis, W.R., 74
Dawson, Elmer, 74
Dawson, Eugene, 84
Dawson, Margaret, 75, 77
DeBoard, Alf, 75
DeRosier, Linda Scott, 3
Dills, Jim, 24
Dills, M/M John, 86
Dorr, Goldwaithe H., 226, 229
Dove, Ernest, 234
Dove, M/M Ernest, 87
Dove, Ruth, 138
Duke University, 155
Dutton, Clarence, 126

E

Eastern Kentucky State College, 61, 126, 140
Elam, Helen, 126
Emmerson, H. B., 222

F

Ferguson, Elanor, 79
Fitch, Archie, 72
Fitch, Delmas, 21
Fitch, Nooner, 21
Fitch, Mrs. Sherman, 84
Fitch, Mrs. Tom, 87
Fletcher, Jess, 213, 215, 216, 225
Fletcher, Grant, 218, 222, 225
Fraley, Billy, 137
Fraley, Don, 183, 198
Fraley, Millard, 76
Franklin, Don, 88
Franklin, Mrs. Eddie, 138
Franklin, Ellen Jean, 138

G

Gambill, Dan, 138
Gambill, Fillmore, 198
Gambill, Naomi, 138
Gibbs, Ruel, 80
Gobel, Dave, 214
Gobel, John W., 214
Golden Rule Hospital, 85
Green, Bernice, 59
Green, Ernie, 58, 59
Green, Jimmy, 28, 58
Green, Libby Ann, 20, 58
Green, Paul, 20, 58
Green, Mrs. Virgil, 84
Green, Virgil, 20, 56
Greene, Albert, 218, 222, 225
Greene, Della Marie, 134
Greene, Ira, 134
Greene, Thelma, 134
Griffith, Mrs. Flem, 79
Griffith, Guthrie Louise, 79
Griffith, Larry, 21
Griffith, Norman, 21

"Grit," 107, 113

H

Hager, Rev. J. S., 80
Hamilton, Myrtle, 78
Hampton, Bill, 181
Hampton, Harry, 199
Hampton, Lizzie, 199
Harper, Mrs. E. J., 240
Harrell, Mrs. Harry, 240
Hayes, Hobert, 79
Hayne, Hugh, 114
Hazelrigg, Mrs. Wm., 133
Heatter, Gabriel, 153, 176
Helton, Will, 214, 219
Hill, Mrs. Jack, 240
Horman, H. S., 53
Horne, Verne P., 132
Howe, Jim, 109
Howes, M/M E. F., 190
Howes, Harry, 53, 70
Howes, Mrs. Harry C., 190
Howes, H. C., 72, 73
Howes, H. S., 221
Howes, Oneida, 189
Howes, M/M Will, 189
H. S. Howes Community School, 34, 65, 111, 123, 132, 136
Huff, Harry, 86, 87, 134
Huffman, M/M W. D., 190
Hughes, W. H., 76
Hunter, Pauline, 134, 135
Hunter, Mrs. Vess, 72
Hyden, Roe, 214

I

Ison, Garnett, 133

J

Jackson, M/M E. B., 86
Jackson, Floyd, 200
Johnson, Mrs. Grant, 81
Johnson, Lady Bird, 142
Johnson, Lyndon B., 119, 142, 143
Johnson, Rhoda, 134

K

Keillor, Garrison, 28
Kennedy, John F., 178
Keyser, Charles M., 237
Keyser Coal Co., 237
Kirk, M. C., 221

L

LaViers, Harry (Henry), 78, 206, 208, 211, 218, 220, 235, 238,
Leek, Lottie, 74
LeMaster, Edna Earl(e), 133, 133, 136
Lewis, A. I., 138
Lewis, Isom, 80
Lewis, John L., 15, 169, 232

Litteral, Sonnie, 85
Lynn, Loretta, 26, 66, 100
Lyons, Rev. Florda, 92
Lyons, Jack Cecil, 21
Lyons, Rev. James, 137, 183, 185
Lyons, M/M James, 87
Lyons, Keith, 20, 87
Lyons, Mitchell, 88
Lyons, Nadine, 86

M

Madison Square Garden, 37
Malone, Buck, 21
Martin, Dr. Robert, 140
May, A. J., 221, 222
May, Jack, 213
Mayo, John C. C. 237
Mayo Memorial Church, 235
Maynard, Rufus, 74
McCloud, Sophia, 82
McCloud, Vivian, 80
McClure, Josephine, 134
McFaddin, Tommy, 80
McKenzie, Carl, 88
McKenzie, M/M Charley, 78, 81
McKenzie, Edward, 134
McKenzie, Ethyl, 80
McKenzie, Marie, 88
McKenzie, Minnie, 80
McKenzie, Nellie, 134

McKenzie, Sue Belle, 134
McKenzie, Walter, 134
Meade, Bud, 48
Meade, Curt, 123, 175
Meade, Emma, 88
Meade, Marshall, 87
Meade Memorial High School, 100, 112, 123, 124, 126, 128, 155, 156
Meade, Mrs. Tollie, 88
Meek, Mrs. J. N., 72
Meek, Warren, 11
Melvin, Roy, 104
Metzger, Carl, 23
Miller, Doris, 20
Miller, Gene, 108
Miller, Mike, 20
Miller, Theodore, 20
Miller, Viola G., 134
Miller, Wanda, 138
Mollett, Burns, 143
Montgomery Ward, 68
Moore, Ed, 215, 225
Moore, Everett, 215, 225
Morris, Mrs. Charles, 240
Mosley, Mrs. Arthur, 88
"Muddy Branch," 90

N

Neal, Mrs. Ernest, 138

Nelson, Glen, 134
Nelson, June Bug, 21
Nelson, Mrs. Vencil, 138
Nelson, Virginia, 134
Nichol, John, 133
Nickels, Wm., 70

O

O'Bryant, Harkless, 3
Old Friendship United Baptist Church, 156
Osborne, M/M Flem, 81

P

Pack, Goldia, 3, 7
Pack, Ernest, 3, 7, 106, 107
Pack, Hubert, 3, 7, 18, 176
Pack, Joe, 3, 5, 7, 87
Pack, Julia, 3, 95
Pack, Lizzie Hampton, 199
Pack, Marcy, 3, 128
Pack, Mary Jean, 3, 7
Pack, Stephen, 199, 201
Pack, Tracy, 80
Pack, Ulysses, 3, 7, 18, 20, 176
Pack, Walt, 130, 156
Pack, Willie, 3, 95
Pack, M/M Willie, 90
Pack, Wilma Jean, 3, 100, 101
Paine, Thomas, 178
Paintsville Furniture Co., 196
Paintsville High School, 112
Pascoe, Samuel, 225, 236, 227, 228, 229
Picklesimer, Virgil, 74, 233, 235
Pelphrey, Ed, 215, 219, 223, 225
Penix, Almira, 140
Penix, Hobert, 140
Penix, Wilma Jean, 140
"Poison Oak," 48
Polo, Marco, 188
"Prairie Home Companion," 29
Preston, Alta Lee, 137
Preston, Bertha Mae, 134
Preston, Bruce, 295
Preston, Callie, 134
Preston, Cecil, 134
Preston, Claude C., 84
Preston, Cyrus, 84
Preston, Delmas, 233
Preston, Edgar, 161, 179
Preston, M/M Edgar, 84
Preston, Ernest Ray, 84, 137
Preston Funeral Home, 92, 128
Preston, George (Bearhunter), 196
Preston, Glen, 69, 72, 189
Preston, M/M Glen, 72
Preston, Mrs. Glen, 72
Preston, Goldia Ray, 88

Preston, Guy Jr., 135
Preston, Jemima Jane, 134
Preston, Mrs. Jim, 81
Preston, Johnny, 233
Preston, McClellan(d), 72, 74, 76,
Preston, Mike, 128
Preston, Paul, 88
Preston, Shady, 80
Preston, Thomas, 134
Prevett, G. G., 214
Prevett, Harry, 214
Prevett, W. T., 214
Price, Charles Lee, 21
Price, Vertrice, 72
Prince, James Harvey, 21
"Progressive Farmer, The," 113
Puckett, Booten, 143
Puckett, Lacy, 194
Puckett, M/M Will, 194
Pugh, Dan, 82

R

Ramey, Russell, 218, 222, 225
Raney, Tom, 53
Ratliff, Fred, 21
Ratliff, Margaret Ann, 21
Ratliff, Milt, 21
Ratliff, Paul, 21
Ratliff, Roger, 21
Red Jacket Rocks, 58

"Reader's Digest," 143
Reynolds, Edgar, 53
Reynolds, George, 20
Reynolds, Georgene, 20
Reynolds, Patsy Grace, 20
Rice, Ernest, 138
Rice, I. G., 221
Rice, Lucy, 72
Rice, Nathan, 80
Rice, Norton, 219, 233, 235
Rice, Walter, 81
Rice, Sterling, 72, 74, 78
Rister, M/M M. G. 75
Robinson, Adaline, 78, 80
Robinson, Emaline, 78, 80
Robinson, Cecil, 134
Robinson, Dorothy, 134
Robinson, Ernest, 200
Robinson, John, 81, 134
Robinson, William, 81, 134
Robinson, M/M Wm., 78
"Rosie the Riveter," 17
Royal Theater, 244
Rucker, Clifford, 72

S

Salyer, Beatrice, 72
Salyers, Ruth, 126
Setser, Jake, 214
Silcott, A. E., 53

Sipp Theater, 244
Shakespeare, William, 64
Sherman, M/M Alfred, 85
Sherman, Fannie, 85
Sherman, Cecil, 134
Sherman, Fred L., 235, 241
Short, Frances Colvin, 137
Short, Ray, 137
Shuey, Mrs. John, 240
Smith, A. Dw., 15, 23, 210, 237
Smith, Alan, 238
Smith, Cove, 214, 216, 219
Smith, Flop, 48
Sneff, Mrs. Earl King, 240
Sparks, Rev. Cully, 92
Sparks, Doll, 21, 92
Sparks, Eck, 21
Sparks, Eskel Lee Castle, 91, 93
Sparks, Faye, 92
Sparks, Jeff, 21, 150
Sparks, M/M Jeff, 91
Sparks, Lizzie, 70
Sparks, Nelson, 70
Sparks, Ray, 92
Sparks, Tom, 92
Sparks, Virginia, 92
Spears, Ella, 81
Spears, Emma, 81
Spears, Ernestine, 137

Spears, Hazel, 80
Spears, Dr. Joe, 3
Spears, Ora D., 233
Spears, M/M P.D., 84
Spears, Sam, 81
Spears, Will, 219, 223, 225
Spencer, Jimmy, 20
Spradlin, Rosy, 81
Stafford, M/M Ray, 72
Staggs, Clyde, 84
Staggs, Charlie, 84
Stambaugh, Garfield, 54
Stapleton, M/M Beecher, 84
Stapleton, Bradley, 196, 201
Stapleton, Cecil, 134
Stapleton, Mrs. Christina, 81
Stapleton, Kathleen, 84
Stapleton, Loman, 134
Stapleton, Olga, 72
Stapleton, Sam, 221
Stapleton, Vance, 134
Stapleton, Virgil, 134
Stapleton, Zelphia, 201
Stimson, Col. Henry L., 226, 229
Stuart, Jesse, 108
Sturgill, Charlie, 134
Sturgill, J. P., 214
Sturgill, M/M Luther, 78
Swayze, John Cameron, 113

T

Taft, Mrs. Alden, 238
Thealka Free Will Baptist Church, 30, 41, 179, 181
Thomas, Charles, 240
Thomas, David, 240
Thomas, Harry, 240
Thomas, Lillian, 72
Thomas, R. C., 211, 235, 239
Thomas, Wildan, 240
Thurman, Roberta, 80
Tom's Creek Quartet, 180
Travis, M/M Estill, 85
Travis, Louisa, 85
Trimble, Johnnie, 21
Trout, Allan, 114
Trout, Wiley, 115
Turner, Dr. John W., 58, 59
Turner, Lida, 198
Twain, Mark, 116

U

"United Mine Workers Journal, The," 107, 113

V

VanHoose, Archie, 234
VanHoose, Bert, 73
VanHoose, Bill, 225
VanHoose, Charlene "Shod," 20
VanHoose, Charles, 76
VanHoose, Claude, 20
VanHoose, Claudine, 20
VanHoose, Earl, 19
VanHoose, Ed, 215, 225
VanHoose, Hubert, 73
VanHoose, James O., 21
VanHoose, Jargo, 19
VanHoose, John Martin, 21
VanHoose, Josephus, 215, 225
VanHoose, Mrs. Leslie, 138
VanHoose, Millard, 199
VanHoose, Oma, 134
VanHoose, Paul, 20
VanHoose, Wilbur, 20
Vaughan, Fred A., 78

W

Wallen, W. M., 233
Ward, George S., 53
Ward, W. J., 213, 221
Ward, Tabor, 75
Webb, Margaret, 3
Weddington, Fanny, 76
Welch, Fanny, 70
Welch, Mary Mae, 134
Welch, Thomas J., 70
Wells, J. H., 214
Wells, W. R., 214
West Point Military Academy, 121

Whalen, Carrol Kathryn, 71
Whalen, M/M Fred, 71
Whalen, Fred Burton, 134
Wheeler, C. B., 221
Wheeler, John W. 221, 225
Wheeler, M. O., 221
White Cloverine Brand Salve, 47
Williams, Ammon, 198
Williams, Carl Thomas, 198
Williams, D. Powell, 199
Williams, Frank, 198
Williams, James Roger, 198
Williams, Linda Bush, 198
Williams, Morris, 23, 236
Williams, O. J., 189
Williams, Patricia Ann, 198
Williams, Ronald, 198
Wyatt, Junior, 80
Wyatt, Ruby, 81

Y

Yeager, Carroll, 74, 77

P. 42-43 Coal dust, gravel roads, and 'THANK YOU LORD for blacktop.